AF759262

Waves of Hope

The U.S. Navy's Response to the Tsunami in Northern Indonesia

Bruce A. Elleman

- 2 APR 2009

NAVAL WAR COLLEGE PRESS
Newport, Rhode Island

UNIV DUNELM

All illustrations are official U.S. Navy photographs unless otherwise indicated.

909.09824083I
ELL

Naval War College

Newport, Rhode Island
Center for Naval Warfare Studies
Newport Paper Twenty-eight
February 2007

President, Naval War College
Rear Adm. Jacob L. Shuford, USN

Provost/Dean of Academics
James F. Giblin, Jr.

Dean of Naval Warfare Studies
Robert C. Rubel

Naval War College Press

Director: Dr. Carnes Lord
Managing Editor: Pelham G. Boyer

Telephone: 401.841.2236
Fax: 401.841.1071
DSN exchange: 948
E-mail: press@nwc.navy.mil
Web: www.nwc.navy.mil/press

Printed in the United States of America

BAS 187

The Newport Papers are extended research projects that the Director, the Dean of Naval Warfare Studies, and the President of the Naval War College consider of particular interest to policy makers, scholars, and analysts.

The views expressed in the Newport Papers are those of the authors and do not necessarily reflect the opinions of the Naval War College or the Department of the Navy.

Correspondence concerning the Newport Papers may be addressed to the Director of the Naval War College Press. To request additional copies, back copies, or subscriptions to the series, please either write the President (Code 32S), Naval War College, 686 Cushing Road, Newport, RI 02841-1207, or contact the Press staff at the telephone, fax, or e-mail addresses given.

Reproduction and printing are subject to the Copyright Act of 1976 and applicable treaties of the United States. This document may be freely reproduced for academic or other noncommercial use; however, it is requested that reproductions credit the author and Newport Papers series and that the Press editorial office be informed. To obtain permission to reproduce this publication for commercial purposes, contact the Press editorial office.

ISSN 1544-6824

ISBN-13 978-1-88473341-3

Contents

Foreword, *by Carnes Lord*		v
Preface		vii
Acknowledgments		ix
CHAPTER ONE	The Tsunami Disaster and the U.S. Political and Military Reaction	1
CHAPTER TWO	Tsunami Warning Delays and Media Reporting Inaccuracies	13
CHAPTER THREE	Preparing the U.S. Navy Operation: The Value of Sea Basing	27
CHAPTER FOUR	Replenishment and Sustainability Issues: Logistics as Leverage	43
CHAPTER FIVE	U.S. Navy On-Site Delivery: Helicopter-Assisted Access	55
CHAPTER SIX	Communications at Sea: Integrating the Collective	69
CHAPTER SEVEN	The Medical Mission: USNS *Mercy* Signals Political Will	79
CHAPTER EIGHT	Wrapping Up the Mission: Lessons Learned and Recommendations	89
CHAPTER NINE	The Political Benefits of UNIFIED ASSISTANCE	101
Afterword: Maritime Strategy and Naval Responses to Nonmilitary Threats		107
Appendix: Chronology of the Tsunami Disaster and Humanitarian Response		119
List of Abbreviations and Acronyms		121
Bibliography		129
About the Author		135
The Newport Papers		137

Foreword

The powerful underwater earthquake that occurred off the coast of Sumatra on 26 December 2004 generated the most destructive tsunami ever recorded, drowning more than 150,000 people without warning in exposed littoral areas from Indonesia to South Africa. The destruction was particularly severe in the Aceh Province of Indonesia, at the northwestern tip of the island of Sumatra. There entire villages were destroyed within minutes as waves of thirty feet or more advanced far inland, while destruction of the main coastal highway made the entire region virtually inaccessible to Indonesian authorities ashore.

In these extraordinary circumstances of human suffering, the U.S. Navy was able to play a key role in organizing what was to become a massive, multinational humanitarian relief operation, one based and executed virtually entirely "from the sea." Working closely with the Indonesian government and military, the Navy delivered, beginning within days of the disaster, vast quantities of emergency food and other supplies and provided on-the-spot emergency medical treatment to thousands of injured and displaced persons along the Aceh coast.

Humanitarian relief has long been recognized as a mission of the American armed forces and of the U.S. Navy in particular. The scale and complexity of the tsunami's impact, however, posed particular and in some respects novel challenges to the Joint Task Force 536 (JTF 536) that was created to deal with the situation, not least of them the requirement imposed on it to operate exclusively from an improvised "sea base," to use a term that has gained some currency in recent discussions of naval missions and capabilities.

In Newport Paper 28, *Waves of Hope: The U.S. Navy's Response to the Tsunami in Northern Indonesia,* historian Bruce A. Elleman provides the first comprehensive history and analysis of what would become known as Operation UNIFIED ASSISTANCE. Elleman, a research professor in the Department of Maritime History at the Naval War College, has produced a valuable and indeed unique study, one that makes use of a variety of internal Navy documents, oral histories, and interviews with a number of senior naval officers, including the then Chief of Naval Operations, Admiral Vern Clark. It is to be hoped that it will prove of immediate benefit to planners in the naval and joint worlds of the U.S. military, as well as to those of other nations potentially interested in exploiting its lessons to improve their own capabilities in this frequently neglected yet vital—indeed, life-saving—military mission.

Thanks are due to many persons in and outside the Navy who have given some of their time to assisting with this project. In particular, I would like to acknowledge the extensive support provided Professor Elleman by the Naval Historical Center in Washington, D.C., as well as the continual encouragement and help of Professor John Hattendorf, chairman of the Maritime History Department at the Naval War College.

CARNES LORD
Director, Naval War College Press
Newport, Rhode Island

Preface

From our own experiences, we know that nothing can take away the grief of those affected by tragedy. We also know that Americans have a history of rising to meet great humanitarian challenges and of providing hope to suffering peoples. As men and women across the devastated region begin to rebuild, we offer our sustained compassion and our generosity, and our assurance that America will be there to help.

GEORGE W. BUSH, 3 JANUARY 2005

The 26 December 2004 Southeast Asian earthquake and the ensuing tsunamis destroyed cities, towns, and huge coastal areas in Indonesia, Sri Lanka, Thailand, India, Malaysia, Myanmar, Bangladesh, the Andaman and Nicobar islands, the Maldives, the Seychelles, Somalia, Tanzania, and Kenya.[1] The U.S. Geological Survey has estimated the human toll at 157,577 people killed, 26,763 missing, and 1,075,350 displaced.[2] In terms of lives lost, it was recorded history's most devastating tsunami.

The day the tsunamis hit, President George W. Bush expressed "his sincere condolences for the terrible loss of life and suffering caused by the earthquake and subsequent tsunamis in the region of the Bay of Bengal."[3] U.S. Pacific Command (PACOM) immediately established an Operations Planning Team (OPT) at PACOM Headquarters at Camp H.M. Smith in Honolulu, Hawaii. On 28 December it formed Joint Task Force (JTF) 536 to plan and execute Operation UNIFIED ASSISTANCE. According to the Pentagon: "The focus of the mission will be to prevent further loss of life and human suffering by expeditiously applying resources to the overall relief effort."[4]

Indonesia was hit hardest; its death toll soon exceeded a hundred thousand people, with the highest mortality rates in northern Sumatra. High mountains cut off Aceh Province from the south, which made the use of sea basing, forward sea-based logistics, helicopter access, communications, and shipboard medical care in the relief effort especially appropriate. The earthquake and subsequent tsunami eliminated the coastal road, thereby cutting off over 110 miles of coast from supply by land;[5] accordingly, air access was crucial to the humanitarian response in Indonesia.[6] Admiral Thomas Fargo, commander of PACOM, emphasized the value of "helicopter vertical lift" to the success of the U.S. Navy's humanitarian mission in Indonesia.[7]

During UNIFIED ASSISTANCE, sea basing proved to be a culturally sensitive and politically flexible response to a natural disaster in a region dominated by Muslims and the scene of an active domestic insurgency. "Hard power" assets, like the aircraft carrier and support ships provided by the U.S. Navy, in conjunction with air support and personnel from the Army, Marine Corps, and Air Force, provided tremendous "soft power" effects. This operation produced enormous goodwill, in particular when compared to the experience of other powers in the region—like China—that could not send similar forces.

Notes

1. The term "tsunami" is composed of the Japanese word *tsu,* which means "harbor," and *nami,* meaning "wave." The often-seen term "tidal wave" refers to astronomical tide waves and so does not adequately describe all forms of seismic events that can lead to impulse wave generation, including earthquakes. In 1963, the term "tsunami" was adopted by the International Union of Geodesy and Geophysics (IUGG) conference to describe these phenomena, and it has remained in general use ever since. The epigraph is quoted from President George W. Bush, "Continuing Support for Tsunami Relief," 3 January 2005, available at www.whitehouse.gov/infocus/tsunami/.
2. U.S. Geological Survey Earthquake Report, "Magnitude 9.1: Off the West Coast of Northern Sumatra," available at earthquake.usgs.gov/eqcenter/eqinthenews/2004/usslav/#summary.
3. Mike Allen, "Flooded Nations Get U.S. Help," *Washington Post,* 27 December 2004.
4. "U.S. Waits to Hear If Troops Needed in Tsunami Area—CNN," *Wall Street Journal,* 28 December 2004.
5. Mark S. Leavitt, Jeffrey M. Vorce, and Michael M. Hsu, "For Compassion and Country: Unified Assistance," U.S. Naval Institute *Proceedings* 131, no. 4 (April 2005), pp. 44–49.
6. Ralph A. Cossa, president of Pacific Forum CSIS, "U.S. Military Provides 'Logistical Backbone' for Tsunami Relief," *eJournal USA: Foreign Policy Agenda* 9, no. 3 (November 2004), usinfo.state.gov/gi/Archive/2005/Mar/05-433015.html.
7. "Navy, Marines Deploy to Southeast Asia for Tsunami Disaster Relief," *Inside the Navy,* 10 January 2005.

Acknowledgments

The author would like to thank the many people who shared their insights and expertise, including Admiral Vernon E. Clark, USN (Ret.), Admiral Thomas B. Fargo, USN (Ret.), Admiral Walter F. Doran, USN (Ret.), and Admiral James A. Lyons, USN (Ret.). At Project Hope, I would like to thank the director, John Howe. At the U.S. Naval War College, I benefited from the advice and support of Mr. Pelham Boyer, Professor John F. Garofano, Professor James Giblin, Professor John B. Hattendorf, Professor Carnes Lord, Captain Ivan T. Luke, USCG (Ret.), Professor S.C.M. Paine, and Professor Robert C. Rubel. Ms. Jo-Ann Parks, Ms. Suzanne Stephenson, and Ms. Margaret A. Richard (of Intekras, Inc.) provided stellar production assistance. At the U.S. Naval War College library, Ms. Alice Juda, Mr. Wayne Rowe, and Mr. Robert Schnare helped me locate appropriate information. At the U.S. Naval War College Security Office, Mr. Leonard Coleman, Ms. Heidi Eldridge, and Ms. Cheryl Vertefeuille assisted me with bringing documents onto campus. At the Naval Historical Center, Rear Admiral Paul E. Tobin, USN (Ret.), the Director of Naval History, gave his enthusiastic support to this worthy project. Commander Doreen Fussman, Ms. Kathy Lloyd, Dr. Edward J. Marolda, Dr. John D. Sherwood, and Dr. Gary E. Weir helped me access their voluminous oral history records, plus Drs. Sherwood and Weir assisted me by reading and giving many helpful comments on how to improve the final manuscript. At the Military Sealift Command, Ms. Beverly McClinton kindly sent me information regarding MSC's participation in the operation. At the Center for Naval Analyses, Captain Peter Swartz, USN (Ret.), and Dr. John S. Ivancovich assisted me. At the Navy Warfare Development Command, Commander Mark A. Henning and Mr. Duane J. Phillips provided me with Navy Lessons Learned reports. Finally, I would like to extend a special thanks to Lieutenant Commander Christopher Hayes and HC-11 for providing photographs.

CHAPTER ONE

The Tsunami Disaster and the U.S. Political and Military Reaction

Words can not adequately describe the physical damage and human suffering brought by this [tsunami], one of the worst natural disasters the world has seen.

ADMIRAL THOMAS FARGO, USN

COMMANDER, U.S. PACIFIC COMMAND, 20 JANUARY 2005

The earthquake that created the 26 December 2004 tsunamis measured just over 9.0 on the Richter scale and was centered under the Indian Ocean floor just to the west of Indonesia.[1] The amount of energy released is thought to have been somewhere between 0.25 and 0.8 gigatons of TNT. Put another way, the U.S. Geological Survey has estimated that the tsunami released energy equal to approximately twenty-three thousand Hiroshima-type atomic bombs.[2]

Tectonic plate shifts displaced tremendous quantities of water along a long underwater rift, creating enormous tsunamis in every direction. Within hours tsunamis had struck the coastlines of over a dozen Indian Ocean countries. Some waves reached fifty feet in height, inundating tourist beaches, destroying businesses and homes, and drowning entire families, including a high percentage of women and children. Hours later, the tsunamis reached eastern Africa, where they produced further devastation and loss of life.

Many relief agencies reported that up to one-third of the dead were children. More girls than boys perished, and as many as four times more women than men died; in Sri Lanka, women and children accounted for many of the thirty thousand deaths. The tsunamis disproportionately impacted beaches, where they killed many foreigners. Early estimates of dead or missing included as many as nine thousand mainly European tourists, who were vacationing at resorts throughout Southeast Asia during one of Europe's peak holiday seasons.

The Location and Timing of the Indonesian Earthquake and Tsunamis

At 7:58 AM local time (00:58:53 UTC) on 26 December 2004, a 9.15-magnitude earthquake struck off the west coast of the Indonesian island of Sumatra. It led to fifteen smaller earthquakes across the region. Altogether these seismic events lasted for ten minutes and produced several massive tsunamis that killed over a hundred thousand people in Indonesia, mainly in the northern province of Aceh (see map 1). These tsunamis were so powerful that they killed one person on Blue Horizon Beach, Port Elizabeth, South Africa, some twelve hours later and over five thousand miles away.

MAP 1
Aceh Province, Humanitarian Reference Map

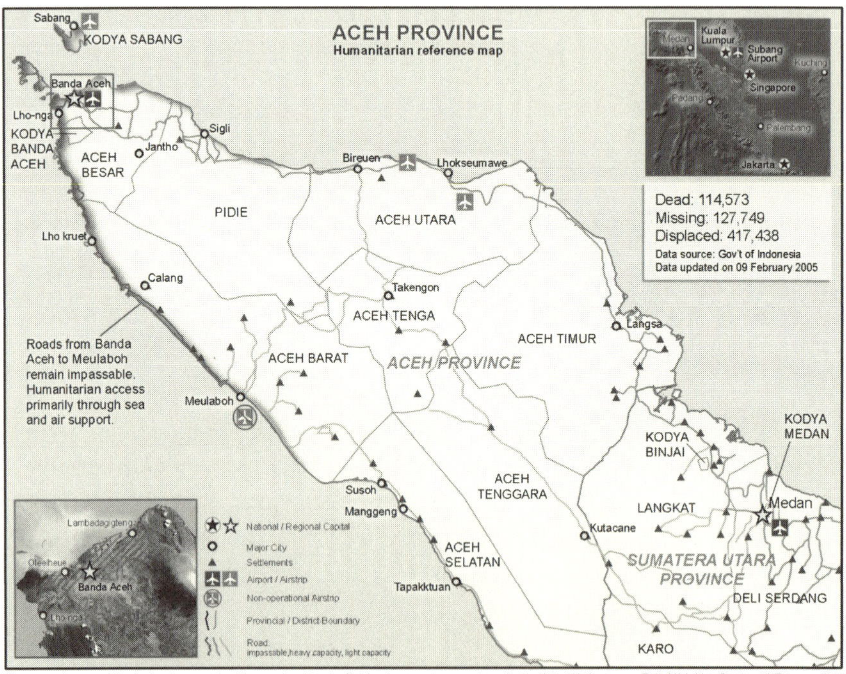

The names shown and the designations used on this map do not imply official endorsement or acceptance by the United Nations ReliefWeb Map Centre - 10 February 2005

According to seismologists, the earthquake's epicenter was approximately a hundred miles west of Sumatra and almost twenty miles below sea level. This point is at the extreme western end of the so-called Ring of Fire, a zone that accounts for about 80 percent of the world's largest earthquakes. Because of the location, reports referred to the "Indonesian" or "Sumatran" earthquake. Others have used the year or the date, calling it the "26 December 2004" earthquake. Because it was still Christmas Day in the United States, some call it the Christmas earthquake, just as many in Britain and many former British colonies refer to it as the "Boxing Day" earthquake.

Tsunami-ravaged coastline, Sawangbatee, Aceh Province, Sumatra, Indonesia.

The earthquake that produced the destructive tsunamis affected some 750 miles of faults where the Indian Plate and the Burma Plate meet. On this occasion these plates slid along each other in two major movements over several minutes. Seismographic and acoustic data show that during the first phase the rupture was about 250 miles long and sixty miles wide, located almost 18.6 miles beneath the seabed; it was the longest plate rupture ever recorded due to a single earthquake.[3] It moved at a speed of 1.7 miles per second, or 6,300 mph, progressing northwest for about two minutes. After a short delay the plates moved again, this time due north, in the direction of the Andaman and Nicobar islands. In addition to plate movement to the northwest and north, the seabed also rose vertically by sixteen feet in some places, displacing an estimated seven cubic miles of water. It was this water displacement, not horizontal movement, that triggered the devastating tsunamis.

The 26 December 2004 earthquake was one of the largest ever recorded. It produced an oscillation of the Earth's surface of eight to twelve inches; the initial shock wave was noticed by seismic stations as far away as Oklahoma, which recorded vertical earth movements of about an eighth of an inch. Overall, the Earth's surface moved vertically by an average of up to one centimeter, or about a third of an inch. But as powerful as the earthquake was, the uniqueness of the tsunamis was related less to the earthquake's size and location than to its exact seismic sequencing. For example, because the second tectonic shift occurred more slowly than the first, at about 4,700 mph, the second set of tsunamis, which traveled to the north, were significantly smaller than those that traveled to the east and to the west.

Even more importantly, the earthquake produced an average thirty-three-foot horizontal movement and a thirteen-to-sixteen-foot vertical movement along the fault line. These shifts changed the geography of the entire region. For example, the Andaman and Nicobar islands seem to have moved thirteen feet toward the southwest, several small islands near Sumatra sixty-six feet, and Sumatra's northern tip as much as 118 feet. Vertical movement was also substantial, creating new hills and depressions both on land and under the ocean. As a result, some coastal areas of Indonesia are now below sea level and may be so permanently.

Unlike many other large tsunamis throughout history, the Sumatran waves did not originate from a single point source but radiated outward along the entire 750-mile length of the fault. The rupture's size, in turn, greatly increased the geographical area over which the waves were observed. Tsunamis reached as far as South Africa to the west, Mexico and Chile to the east, and the Arctic Ocean to the north. Sensors recorded the tsunami all over the world, from the North Pacific to Antarctica, including most of the recording sites in the Kuriles, the Aleutians, Alaska, British Columbia, California, Mexico, Peru, and Chile.[4]

The height of the individual tsunamis differed radically from area to area, depending on the direction the shoreline faced and the depths of the surrounding waters. For example, along Sumatra's northwestern coastline some waves were over thirty feet high, while in Sri Lanka and Thailand the average wave height reached twelve to fifteen feet. While most of the worst wave damage occurred in areas closest to the seismic ruptures, this was not always the case. Because of the east-west direction of the underwater movements, some waves that hit the Seychelle Islands and Somalia, on the far side of India, reached approximately sixteen feet in height. A full sixteen hours after the earthquake, a final five-foot wave crashed into Struisbaai, South Africa, over 5,300 miles away from the earthquake's epicenter.[5]

Tsunamis throughout History

Although tsunamis are perceived as being extremely rare, they actually appear regularly in all parts of the globe, including the Indian, Pacific, and Atlantic oceans. During the past century alone, many large ocean tsunamis were triggered by earthquakes near Indonesia and Pakistan, and in the Bay of Bengal. The review that follows of eight major tsunamis during the past two and a half centuries gives some idea of their potential destructiveness.[6] In most of these cases, coastal populations were caught unaware, which increased the impact enormously.

On 1 November 1755, one of the largest earthquakes reported to that time occurred in the Atlantic Ocean just off the coast of Portugal. The earthquake reportedly lasted ten minutes and has since been estimated at 9.0 on the Richter scale. The tremors themselves destroyed most of downtown Lisbon. About thirty minutes later, an enormous tsunami struck Lisbon's harbor and traveled far into the city. Tsunamis later hit the coasts of Spain, France, Belgium, Holland, England, and Ireland. It is estimated that over a hundred thousand people perished in all. In Lisbon, over one-third of the population reportedly died.

On 26 August 1883, Krakatoa's volcano exploded; the island quickly sank into the ocean, triggering large tsunamis, some over 120 feet in height. Since Krakatoa was uninhabited, it is thought that no one died in the explosion, but the resulting tsunamis reportedly killed over thirty-six thousand people. The islands of Java and Sumatra were struck the hardest. A number of towns, including Telok Batong on Sumatra and Sirik and Semarang on Java, were completely destroyed. One heavily populated area on the southwestern tip of Java was never resettled; to this day the Ujung Kulon region of Java is part of the Krakatoa nature preserve. Java and Sumatra sustained most of the damage, but tsunamis were reported in the Indian Ocean, the Pacific Ocean, along the American west coast, and in South America. Higher than average waves were reported even in the English Channel.

On 15 June 1896, a Richter 7.2 earthquake centered off the Japanese coast near Sanriku triggered a massive tsunami that devastated the city and killed over twenty-six thousand people. The height of some tsunamis reportedly reached eighty feet. Tsunamis were also observed across the Pacific Ocean as far away as California; on 16 June 1896 the *San Francisco Chronicle* reported ten-foot-high waves.

On 28 December 1908, an earthquake created a tsunami that struck Messina, Sicily, and the province of Reggio di Calabria in southern Italy. Estimates of the dead ranged as high as two hundred thousand. Coincidentally, the U.S. Navy's "Great White Fleet" was passing through the Mediterranean at the time. Having already responded to the 1906 San Francisco earthquake, the fleet now rushed to the scene of the disaster to provide aid. Experts later disagreed about whether this humanitarian assistance really made a difference, but the mayor of Naples complimented the Navy "for the tactful manner" in which supplies had been placed at the disposal of the Italian government.[7]

On 22 May 1960, a 9.5 earthquake, widely considered the largest recorded earthquake in history, struck Chile. Earthquake-triggered tsunamis damaged the Chilean coast, killing some two thousand people. Tsunamis also spread out across the Pacific Ocean. Fifteen hours later, thirty-foot-high waves hit Hilo, on the island of Hawaii, resulting in sixty-one deaths. Twenty-two hours after the earthquake, ten-foot tsunamis struck Japan, causing two hundred deaths. Tsunamis struck as far away as the Marquesas, in Samoa, and in New Zealand.

An enormous tsunami popularly called the "Good Friday" tsunami hit the west coast of North America on 28 March 1964, following a 9.2 earthquake centered near College Fjord in Prince William Sound, Alaska. This earthquake lasted almost five minutes; the resulting thirty-eight-foot-high tsunami swamped the small Alaskan coastal communities of Girdwood, Portage, and Valdez, killing 106 and making this tsunami the most devastating ever to hit the continent of North America. Tsunamis also affected British Columbia and the west coast of the United States; in Crescent City, California, eleven people perished in high waves. Tsunamis also caused damage in the Hawaiian Islands.

On 16 August 1976, a 7.6 earthquake struck in the Moro Gulf, in the Philippines, only a few miles from Mindanao Island. The earthquake created widespread damage, and tsunamis engulfed numerous coastal communities of the Sulu Archipelago and southern Mindanao, including Zamboanga City and Pagadian City. The tsunami inundated four hundred miles of coastline bordering Moro Gulf in the North Celebes Sea, with over five thousand people killed immediately and thousands more reported missing.

On 17 July 1998, a 7.1 earthquake centered about fifteen miles from the coast of northern Papua New Guinea triggered an undersea landslide that in turn created tsunamis reaching forty feet high. An enormous tsunami hit the Papua New Guinea coastline only ten minutes later, destroying the villages of Arop and Warapu. An estimated 2,200 people died.

These historical cases show that devastating tsunamis are in fact not rare, especially in the Pacific Ocean. The U.S. National Oceanic and Atmospheric Administration estimates that there are an average of two tsunamis each year that cause damage to people or property and that every ten to twelve years a tsunami causes widespread death and destruction.[8] The 26 December 2004 tsunamis in Southeast Asia were just such an event.

Tsunami Destruction in 2004

The widespread destruction associated with the Indonesian earthquake was the result not just of the earthquake per se but of the enormous tsunamis it created. These waves radiated outward in all directions from the rupture. As we have seen, because the seismic events were sequenced, so too were the tsunamis. Some areas received not just one but multiple waves, separated by several minutes. Many people who survived a first wave assumed that the worst had passed, only to be swept away by a second, often larger, wave that arrived a few minutes later.

High mortality rates were exacerbated by the low level of tsunami awareness among the population. In many low-lying areas people did not respond adequately either to the earthquake or to the arrival of the first tsunami. There were some exceptions, however, particularly in rural communities: "In a few cases, particularly in Indonesia and Thailand, isolated communities had retained an ancestral memory of similar disasters and had fled to higher grounds when alerted by the initial tremors."[9]

Many relief agencies reported that up to one-third of the dead appear to have been children, some of whom had been attracted toward the water by the unusual expanse of sea bottom exposed by the lower than average intertidal zone (the area between low and high tides) that preceded the waves. Overall, many more girls than boys seem to have perished. Several factors could explain this disproportion. First, more girls would have been trapped indoors, since the waves hit just around

Tsunami damage at Pasi, northwest coastline of Aceh Province.

8 AM in Indonesia and just after 10 AM local time in both Thailand and Sri Lanka; more boys than girls were probably outside playing.[10] Second, boys were more likely than girls to react to the arrival of the waves by running away, climbing trees, or scaling the outside of buildings. Finally, more boys than girls probably knew how to swim and so had a better chance of reaching safety after the waves hit.

Gender played a role too among adults. In some locales as many as four times more women than men were killed; in Sri Lanka it is thought that women and children accounted for the majority of the over thirty thousand tsunami-related deaths.[11] Such differences may be attributable to the fact that many of the men were outside working, either on the water as fishermen (the tsunamis would have rolled harmlessly beneath their boats) or in urban areas farther from the coast and where concrete buildings could better withstand the destruction. By contrast, many women were caught inside their homes when the waves struck.

In some countries, for example Sri Lanka, women do not normally learn how to swim or climb trees. Also, their long hair and saris made it harder for them to escape the rising waters. According to one account by a young girl who watched her thirty-six-year-old mother drown, "The water came with a huge force, moving like an angry monster across the sand and into the home. My mother helped my younger brother to tear off [*sic*] his shorts to swim away, but she didn't follow. She was just too modest to remove her clothes to escape."[12] Many other women hesitated to leave their homes unescorted by men and so refused to leave and seek higher ground after the first tsunami hit; a second wave swept them away a few minutes later.

As for the many foreign tourists killed by the tsunamis, Scandinavian countries appear to have been hardest hit, with Sweden alone accounting for over five hundred of the more than 2,400 foreign deaths.[13] American casualties, by contrast, were small. From an original number, in early January 2005, of four thousand Americans missing, by mid-February the State Department was reporting that only thirty-three Americans had died or were presumed dead as a result of the tsunamis.[14]

Measured in lives lost, the Indonesian earthquake and tsunamis collectively qualify as one of the ten worst disasters in recorded history. In terms of dead and missing together, however, the tsunamis alone represent the single worst *tsunami* event, specifically, in known history. It is important not to equate the two categories. The press has widely misreported the 26 December 2004 tsunamis as the world's worst disaster ever, an incorrect impression that may have been prompted by the high numbers of Western tourists involved, the presence of radio and television stations near the disaster sites, and the high speed at which footage from the disaster reached the Internet and global television.

Actually, many nineteenth- and twentieth-century disasters killed far more people. For example, in 1887, China's Yellow River broke through its banks and killed an estimated two million, and then again in 1931 and 1938, producing four million and one million more estimated deaths, respectively.[15] In 1970, over three hundred thousand people were killed in Bangladesh when a cyclone with winds of 130 mph swept away entire villages.[16] Finally, in 1976, an earthquake located in Tangshan, China, killed anywhere from 242,000 to 750,000 people, with 665,000 a widely accepted estimate.[17] So, while the December 2004 event was unquestionably the world's worst recorded tsunami, it was far from being the world's worst recorded disaster of all time.

The American Response

When the news of the Sumatran earthquake and tsunamis reached PACOM headquarters at Pearl Harbor, Hawaii, its commander, Admiral Thomas Fargo, set up an operations planning team. On 26 December the international press reported slightly over ten thousand casualties throughout the entire region, a figure suggesting that a small humanitarian assistance/disaster relief operation might be sufficient. Over the next three days, however, PACOM constantly updated and expanded the scope of what became known as Operation UNIFIED ASSISTANCE.

PACOM was able to respond quickly, since just a few months earlier the Joint Intelligence Center Pacific (JICPAC) had created the Contingencies Operational Intelligence Cell, a "fully manned, all-source operational intelligence capability specifically structured to respond quickly to emerging crises within the theater."[18] On 28 December PACOM established Joint Task Force (JTF) 536 under the command of Lieutenant General Robert R. Blackman, Jr., commanding general of the 3rd Marine Expeditionary Force. PACOM changed the task force's name on 3 January 2005 to Combined Support Force (CSF) 536, to reflect the multinational nature of an operation that by then included not only U.S. military forces assisting Indonesia, Sri Lanka, and Thailand but also forces from a coalition of Australia, Japan, Singapore, Russia, France, and Malaysia.

By mid-January 2005 Australia had more than nine hundred troops in Aceh Province; Japan had deployed two ships with 350 troops; and Singapore had sent the tank landing ship *Endurance,* supported by helicopters, as well as troops; while Germany, Britain, and China had each sent medical teams.[19] The Australian Defence Force began its humanitarian response on 28 December 2004;[20] on that day the first Australian C-130 transport aircraft landed at Banda Aceh airport.[21] The first Singaporean relief flight to Aceh Province arrived on 30 December 2004, when a Republic of Singapore Armed Forces Chinook helicopter delivered water, food, and medicine to the people of Meulaboh; *Endurance* arrived on 2 January 2005.[22]

Lieutenant General Blackman first stood up JTF 536 at Utapao, Thailand, at the Royal Thai Navy Base, known as "Camp Red Horse." Blackman readily acknowledged that he had "really very little information on the extent of the disaster and the unique requirements of each of the three primary countries that we're conducting relief operations in."[23] Utapao, however, was an obvious choice for his headquarters, since it had been the focus of numerous combined U.S.-Thai exercises in the COBRA GOLD series, a fact that allowed U.S. forces to "come in here to Utapao at a tremendous comfort level";[24] the advance team for that year's COBRA GOLD, usually held in May, simply needed to speed up its arrival.[25] Blackman's choice facilitated a rapid military response also "because of the expeditionary forces and forward-based forces here in the region."[26]

On 28 December the first elements of JTF 536 began to arrive in Utapao to set up the Forward Command Element. By the 29th the first disaster relief assessment teams, designated DRAT-Thailand and DRAT–Sri Lanka, had already arrived in their respective countries, while DRAT-Indonesia did so the following day. Also on 29 December, the first P-3 reconnaissance flight took off, and the first C-130 arrived at Utapao. On 30 December, the first C-130 relief flight brought supplies into the region, while by 31 December the first helicopter relief flight was operational.

On 2 January 2005, Lieutenant General Blackman and the main body of his JTF 536 team arrived in Utapao to take charge of the operation, followed by the arrival of PACOM's Multinational Planning Augmentation Team. MPAT specialists trained in international disaster response helped set up JTF 536's Combined Coordination Center. Soon they began to receive updates from JICPAC. The "interagency partnership between the theater intelligence center and national intelligence agencies ultimately resulted in the highly successful delivery of information to forward-deployed forces."[27]

If C-130s were in the air first, by 5 January 2005 the Air Mobility Command, a component of U.S. Transportation Command, was supporting the operation with six larger C-5 aircraft flying from Kadena Air Base in Japan and four McChord Air Force Base C-17s staged out of Utapao. Another C-17, from Charleston Air Force Base, South

The Military Sealift Command hospital ship USNS *Mercy* operating with the USS *Abraham Lincoln* on station near Banda Aceh.

Carolina, departed for the region on 4 January 2005.[28] Unlike the Vietnam War, when the runways at Utapao "rumbled and roared with the takeoffs of massive B-52 bombers heading east on bombing runs," this mission soon ran "on cell phones, e-mail, Web pages and paperless offices."[29]

On 29 December 2004, President George W. Bush pledged $35 million in relief assistance; by 31 December he had increased this commitment to $350 million. Land forces were to assist Thailand and Sri Lanka; the U.S. Navy's primary mission was in northern Indonesia. A strike group led by the carrier USS *Abraham Lincoln* (CVN 72) quickly left Hong Kong, and a seven-ship expeditionary strike group led by the helicopter/dock landing ship USS *Bonhomme Richard* (LHD 6) proceeded from Guam. On 1 January 2005, the U.S. Department of Defense (DoD) activated the hospital ship USNS *Mercy,* which departed San Diego on 8 January. By 5 January 2005, only ten days after the earthquake and tsunamis, UNIFIED ASSISTANCE included over twenty-five U.S. Navy ships, forty-five fixed-wing aircraft, and fifty-eight helicopters. By this point American forces had already delivered more than "610,000 pounds of water, food and other supplies to the region."[30]

* * * * * * *

While the 26 December 2004 disaster was not the worst ever in world history, it produced the single worst tsunami crisis ever recorded; the resulting catastrophe certainly qualifies as one of the ten worst in recorded history. Perhaps due in part to the foreign presence, the global awareness of this disaster was both rapid and extensive. As will be seen, within minutes e-mails reported the tsunamis, and within hours the Internet was full of pictures and even home movies showing the widespread destruction. This in turn would prompt the United Nations to establish a relief operation, which the UN has since declared to be the costliest ever undertaken. The tsunami hit Indonesia particularly hard; President Susilo Bambang Yudhoyono would tell Secretary-General Kofi Annan that full recovery might take parts of his country "five to ten years."[31]

By chance, many U.S. Navy Force Elements were already in East Asia, and these ships almost immediately were diverted to the disaster area. But it took several days for the first of them to arrive from Hong Kong and many more days for other naval assets to travel from Guam, Hawaii, and the continental United States to Southeast Asia. Nonetheless, by 5 January 2005, only ten days after the earthquake and tsunamis, 13,435 American military personnel were involved in UNIFIED ASSISTANCE. To support this effort, PACOM assigned the *Abraham Lincoln* and *Bonhomme Richard* groups.

The American response is particularly striking in that during the first day or two after the disaster most news reports did not indicate the true severity of the damage or even hint at the large number of casualties. Estimates of dead and missing ranged only in the low thousands, up to a "high" of just over ten thousand. The local destruction had in fact been so great that only two or three days later did the true extent of the catastrophe start to become apparent.

Notes

1. Epigraph from Admiral Thomas Fargo, Commander, U.S. Pacific Command, Media Roundtable Transcript, Kuala Lumpur, Malaysia, 20 January 2005, available at www.pacom.mil/speeches/sst2005/050119transcript.shtml.
2. "The Deadliest Tsunami in History?" *National Geographic News,* 7 January 2005, available at news.nationalgeographic.com/news/2004/12/1227_041226_tsunami.html.
3. National Science Foundation, "Analysis of the Sumatra-Andaman Earthquake Reveals Longest Fault Rupture Ever," 19 May 2005, available at www.nsf.gov/news/news_summ.jsp?cntn_id=104179.
4. Fisheries and Oceans, Canada—Pacific Region, "Tsunamis and Tsunami Research: December 26, 2004 and March 28, 2005 Indian Ocean Tsunami Events," www.pac.dfo-mpo.gc.ca/sci/osap/projects/tsunami/tsunamiasia_e.htm.
5. Ashfak Mohamed, "Freak Waves Hit SA as a Result of the Tsunami," *Cape Times,* 28 December 2004.
6. For more information on these and other tsunamis see "Tsunami History of Tidal Waves," www.important.ca/tsunami_history.html.
7. James R. Reckner, *Teddy Roosevelt's Great White Fleet* (Annapolis, Md.: Naval Institute Press, 1988), p. 148.
8. "Physics of Tsunamis," National Oceanic and Atmospheric Administration, wcatwc.arh.noaa.gov.
9. United Nations, *Report and Summary of Main Conclusions,* Regional Workshop on Lessons Learned and Best Practices in the Response to the Indian Ocean Tsunami, Medan, Indonesia, 13–14 June 2005 (New York: Office for the Coordination of Humanitarian Affairs, 5 July 2005), available at www.reliefnet.int.
10. Sri Lanka is an hour behind Thailand, but it was also farther away from the epicenter, so

the two countries experienced tsunamis at the same time locally.

11. "UNICEF Says 12,000 Children Died in Sri Lanka," CBC News, 2 January 2005, at www.cbc.ca/world/story/2005/01/02/srilanka050102.html; Sarala Emmanuel, "Sri Lankan Women's Small but Significant Gains in the Post Tsunami Reconstruction Process," Suriya Women's Development Centre, Sri Lanka, n.d., at www.apwld.org/tsunami_reconstruction.htm.

12. Suvendrini Kakuchi, "In Tsunami, Women Put Modesty over Survival," Womensenews.com, 23 February 2005.

13. "Swedes Mark Tsunami Anniversary," BBC News, 26 December 2005, available at news.bbc.co.uk/1/hi/world/europe/4560468.stm.

14. Glenn Kessler, "White House to Seek $950 Million for Tsunami Aid," *Washington Post*, 10 February 2005.

15. "Dealing with the Deluge," *Nova*, available at www.pbs.org/wgbh/nova/flood/deluge.html.

16. "Crisis Prevention and Recovery: Bangladesh," United Nations Development Program, www.unpd.org.

17. "1976: Chinese Earthquake Kills Hundreds of Thousands," BBC News, available at news.bbc.co.uk/.

18. David J. Dorsett, "Tsunami: Information Sharing in the Wake of Disaster," *Joint Forces Quarterly* 39, available at www.dtic.mil/doctrine/jel/jfq_pubs/0539.pdf.

19. Denis D. Gray, "Indonesia Denies U.S. Pilots Use of Airspace; Carrier Leaves," *Washington Times*, 13 January 2005.

20. "Indonesia: Operation Sumatra Assist," www.defence.gov.au/our_people/op_sumatra_assist/index.cfm.

21. Brigadier Dave Chalmers, "Operation SUMATRA ASSIST: A Command Perspective," *Defence* (November/December 2005), available at www.defence.gov.au/defencemagazine/.

22. "Case Study: Singapore Armed Forces (SAF)," in *Tsunami: Magnitude of Terror*, library.thinkquest.org/04oct/01724/relief_saf.html.

23. Lieutenant General Robert R. Blackman, commander of Combined Support Force 536, "DoD Briefing on Operation UNIFIED ASSISTANCE, the Post-Tsunami Relief Effort," Utapao, Thailand, 7 January 2005, available at www.pacom.mil/speeches/sst2005/050107tsunami.shtml.

24. Kathleen T. Rhem, "'Conditions for Transition' Met in Thailand, Sri Lanka," American Forces Press Service, 15 January 2005.

25. Matt Pottinger and Barry Wain, "U.S. Military Ties to Region Allowed Faster Response," *Asian Wall Street Journal*, 10 January 2005.

26. Blackman, "DoD Briefing."

27. Dorsett, "Tsunami."

28. "AMC Aircraft, People Support Tsunami Relief Operations," *Defense Transportation News* 9, no. 1 (January 2005), available at www.ndtahq.com/download/0105NDTAGRAM.pdf.

29. James Brooke, "Ad-hoc Planning Goes into Marine Aid Effort," *New York Times*, 6 January 2005.

30. Jim Garamone, "More Ships, Helicopters Arrive to Aid Tsunami Victims," American Forces Information Service, 5 January 2005, available at www.defenselink.mil/news/Jan2005/n01052005_2005010509.html.

31. "Tsunami Recovery Could Take Ten Years, Will Cost Billions," Environment News Service, 7 January 2005, available at www.ens-newswire.com/ens/jan2005/2005-01-07-02.asp.

CHAPTER TWO

Tsunami Warning Delays and Media Reporting Inaccuracies

I've been in war, and I've been through a number of hurricanes, tornadoes and other relief operations, but I have never seen anything like this.

COLIN POWELL, U.S. SECRETARY OF STATE, 6 JANUARY 2005

Soon after the 26 December 2004 tsunami struck, news of the disaster began to trickle out in the international press and over the Internet.[1] On 27 December 2004, official estimates of the dead were comparatively low, with some reports suggesting upward of 13,500 people. These reports were provisional, however, and were expected to rise as relief and rescue workers combed through the wreckage.

The December 2004 tsunami was the first major one to hit Southeast Asia or the subcontinent in over a hundred years, since the tsunami caused by the 1883 explosion of Krakatoa. Many people were simply not aware that an earthquake could create a lethal tsunami. Thousands of lives could have been saved if local governments had taken immediate action to alert the most vulnerable areas to the possibility of seismically created waves. However, because there was no regional tsunami disaster system in place, the governments of these endangered countries were unaware that lethal tsunamis might be heading their way.

After the tsunamis struck, the same poor communications that had contributed to the disaster conditioned the underestimations of the true scale of the damage. Unreliable communications meant that initial death tolls were vastly underreported and, accordingly, that foreign aid did not begin to flow to the region as quickly as it might have otherwise. Only belatedly did the governments in the affected areas announce that they were unable to cope by themselves, and only two or three days after the tsunamis struck did the international press begin to understand and report the true scale of the disaster.

Tsunami Warning

The Pacific Ocean region has the Pacific Tsunami Warning Center (PTWC) at Ewa Beach, Hawaii, providing tsunami warnings to most countries in the Pacific Basin; there is no similar warning system for the Indian Ocean. Dr. R. S. Dattatrayam, director of seismology at the India Meteorological Department, later explained to the *Times of India* that his team had detected the quake but could not predict the tidal wave: "We had indications pretty early in the morning, almost soon after it [the tsunami] originated [in Indonesia]. But we were not prepared to gauge it. We don't have warning facilities for tsunami. We knew something would be hitting us, but couldn't tell the time, the location and the intensity."[2]

Such limitations affected not just the local countries but the PTWC and the Pacific Ocean as well. Following the earthquake, Charles McCreery, director of PTWC, knowing that the earthquake might create tsunamis, transmitted warnings to the U.S. Navy and U.S. State Department, the government of Australia, and the American military base on the British-controlled island of Diego Garcia.[3] The State Department reportedly passed it on to India, but Indian government officials later denied that they had received any such warning, leading some to suspect a cover-up.[4]

But the tsunami center in Hawaii could not get warnings through to the Southeast Asian governments. Professor Michel Chossudovsky, director of a globalization institute at Ottawa University, later argued that with modern communications "the information of an impending disaster could have been sent round the world in a matter of minutes, by email, by telephone, by fax, not to mention by satellite television."[5] However, even if a tsunami warning had been pushed downward into the region, there was no way of ensuring that it would reach the right people. In fact, any such warning might have easily been diverted to the wrong people or ignored completely as a hoax.

The main problem was that there was no regional tsunami disaster system in place. Even had individual countries issued their own tsunami warnings, there was no guarantee that the average person would have heard of them in time. In the case of Indonesia, the total time available would have been only fifteen to thirty minutes, clearly insufficient to evacuate so many people. So, with no widespread and comprehensive warning system in place, most of the population could not have been notified of the approaching waves.

To help remedy this situation, on 29 December 2004 President Bush declared American support for the construction of a worldwide system for such natural disasters. On 9 February 2005, Bush pledged $35 million to help build a tsunami early warning system in the Indian Ocean, earmarking $23 million to improve the international and U.S. tsunami early warning system, and $12 million to enhance tsunami early warning and disaster mitigation in the affected countries. The Southeast Asian system was to be modeled on the 1949 system built by the United States, Japan, and other Pacific countries; it would include a secure communication network to notify the countries of Southeast Asia of an impending disaster from the sea.

The First Stage of International Reporting

The tsunami disaster and the relief effort it generated show how the international press has become an integral factor in understanding and responding to natural disasters. The size and scope of the disaster were at first unclear in part because the global media was not represented in the most affected regions. As a result, the international response to the earthquake and resulting tsunamis was similarly sporadic and uncoordinated. It would take several days, and in some areas almost a week, before a true picture of the devastation was available.

In the wake of the tsunamis, local relief and rescue efforts, limited to begin with, were quickly overwhelmed. Local governments were simply not prepared for a disaster of such proportions. Emergency workers were hampered by shortages of equipment and supplies, as well as by poor transportation. Many major roads, railroads, and docks had been destroyed by the earthquake and tsunamis. In Indonesia, President Susilo Bambang Yudhoyono quickly declared a national disaster and promised aid to the victims, but there was no way to deliver it.

As was generally the case, very few international organizations understood during the first day or two the actual magnitude of the disaster or how many people had perished. Even though Indonesia was the primary focus of the earthquake and tsunamis, the Australian Broadcasting Corporation (ABC) initially put the official death toll at only 4,500, and identified most of the victims as coming from Aceh Province, at the northern tip of the island of Sumatra. Of this number, at least three thousand of the deaths, or almost three-quarters, were reported to have been in the provincial capital of Banda Aceh. In fact, throughout northern Sumatra entire villages had been wiped out. Initially, Australian Broadcasting Company news could only speculate about widespread reports of collapsed buildings in the area; it acknowledged that there were no reports from Aceh's west coast, closest to the earthquake's center, where damage would probably have been most severe.[6]

Other countries in South and Southeast Asia were also cautious in their casualty estimates. Early reports from Thailand stated that at least 310 people had been killed, including victims at several southern tourist resorts, such as the island of Phuket. More than 1,300 others had been injured in six provinces. "We are in chaos," Somsak Sunwansujarit, deputy director of the county's disaster department, admitted.[7] In

Flooded inland plain near Keude Panga, Aceh Province.

Malaysia, at least fifty-three people were immediately reported dead and another thirty-four were reported missing after fifteen-foot waves struck the coastline. The government of the Maldives declared a state of emergency after the tsunami swept over the islands and flooded two-thirds of the capital. However, it placed the death toll at only fifteen. As later events were to show, such estimates were far below the actual numbers.

The Sri Lankan defense ministry quickly confirmed the deaths of 2,484 people, but unofficial estimates offered a higher figure, around five thousand. For example, the initial death toll for Amparai was nine hundred, for Batticaloa three hundred, and for Trincomalee three hundred. But residents of another village, Saindamarudu, were saying that the true number was much higher, anywhere between three and four thousand. The capital, Colombo, lies on Sri Lanka's west coast, away from the earthquake, but the tsunamis "ricocheted" off the east coast of India and hit Colombo hard, inundating as well a number of coastal areas nearby. President Chandrika Kumaratunga returned from a trip to Britain, quickly declared a "state of disaster," authorized the use of the military, and appealed for international aid.[8]

The Sri Lankan government eventually estimated that over a million people, perhaps 5 percent of the entire population, had been directly affected by the disaster. Hundreds of international tourists had also been affected. The seventy-four-year-old former German chancellor Helmut Kohl was in Sri Lanka at that time. Kohl watched from his third-floor balcony as the tsunami crashed into his hotel: "The sea had swallowed up everything. Visions of the war which I experienced as a boy came to mind. It looked like it used to be after a bombing raid. . . . We saw dead people." Kohl elected to remain in Sri Lanka to help with the relief efforts.[9]

Early reports from India stated that at least three thousand people had been killed in southern India, the state of Tamil Nadu accounting for over 1,700 dead. In the state capital of Chennai the official toll stood at 131, but the newspaper *Hindu* reported that hundreds of fishing boats had been destroyed or washed out to sea, which suggested a larger disaster.[10] According to the Associated Press, P. Ramananmurthy in the town of Kakinada explained: "I was shocked to see innumerable fishing boats flying on the shoulder of the waves, going back and forth as if made of paper. Many boats were upturned, but fishermen were still holding onto them. They also were pushed into the sea. I had never imagined anything like this could happen."[11]

In the meantime, the fates of the Andaman and Nicobar islands were still unknown. Because of their location and the extreme damage they had suffered, local and foreign reporters could not get there and communications were cut off. These islands had not only been directly in the path of the tsunami but had also been shaken by a major aftershock. As S. B. Deol, inspector general of police, told the Indian press: "The situation is very grim. The death toll will go up to at least 1,000."[12] As later reports would reveal, even this seemingly large number was a gross underestimate.

International Reactions

News of the earthquake and tsunamis first appeared in the United States on Sunday, the day after Christmas. Estimates of just over ten thousand deaths throughout the region were common, and the international press's underreporting on this day in particular of the death and destruction was to have serious repercussions. Early reports, placing the number of dead per country in the low thousands instead of the tens of thousands—awful enough but a tenth or even a fifteenth of the true number—muted the international reaction to the crisis.

The main problem appears to have been that seasoned international reporters went out of their way not to exaggerate their estimates, reporting thousands of dead when the real figure was well over a hundred thousand for Indonesia alone. Undoubtedly, and understandably, these journalists were concerned that overreporting casualties might lead to accusations of sensationalism. But their reaction gave a falsely reassuring impression to the rest of the world.

Given the comparatively low mortality estimates in the international press, it should not be surprising that the initial reaction of official Washington, D.C., which was in its Christmas to New Year's break, was at first apparently lackadaisical. Also, although it was Saturday, 25 December 2004, in Washington when the earthquake struck, news began to trickle in only on Sunday—and of all Sundays of the year, perhaps the worst in which to find anyone on duty. In fact, most of the main decision makers in the White House,

Congress, and the Pentagon were away from their offices. Many government officials, like President Bush, who was spending the Christmas vacation at his ranch in Texas, were not even in Washington. Reliance on press reports, then, for accurate estimates of the extent of the damage was probably greater than normal. As Senator Sam Brownback (R-Kan.) commented, because of the timing of the disaster "there [was] just not the level of communication you would normally have on something like this."[13]

Nonetheless, Admirals Thomas Fargo, commander of the U.S. Pacific Command, and Walter Doran, commanding the Pacific Fleet, both in Honolulu, were quickly authorizing the movement of ships toward the disaster scene. The fact that Washington decision making had ground to a standstill over the holidays proved on this occasion a boon for Admirals Fargo and Doran, for nobody was trying to look over their shoulders and second-guess their decisions. Though not yet sure which countries had sustained the most damage, they ordered all available naval assets to begin to flow toward Southeast Asia.[14]

As news reports about the tsunami began to appear, starting on the Internet, with graphic pictures of the destruction, and a day or two later in the world press, Pacific Command planners scrambled to overcome the widespread misapprehension that this disaster was isolated geographically and relatively small in scope. Then the first news

Commander, U.S. Pacific Command, Admiral Thomas Fargo, Pearl Harbor, Hawaii

Commander, U.S. Pacific Fleet, Admiral Walter F. Doran, speaking to members of the Institute for Continued Learning at the University of California in San Diego, Calif.

reports came out of Aceh Province, and it finally became clear that Indonesia was the true focus of the catastrophe.

The Second Stage of Reporting

By 27 December more reliable estimates of the earthquake and tsunami-related destruction were available. For example, the official death toll in Aceh Province jumped from around four thousand to over seven thousand people, and Vice President Joseph Kalla was warning that it might soon exceed twenty-five thousand. But news reports from northern Indonesia were still largely confined to Banda Aceh. Because of widespread infrastructure damage, it took two days for Indonesian military aircraft to survey the Sumatran west coast. What they discovered was widespread earthquake and tsunami damage affecting some five million people. According to one estimate, 1,556 of Aceh Province's 5,862 villages, over a quarter, had been completely destroyed.[15] Most of these much more negative reports only began to circulate a day or two after the tsunamis, mostly from the 28th and well into the New Year. On 29 December, however, the situation in Aceh still "remained a mystery, U.N. officials said."[16]

In Thailand the official death toll stood at 1,520, but at least 1,400 others were reported missing. The hardest-hit areas were the seaside resorts of Phuket and Phang Nga, where many foreign tourists had perished. Media reports showed chaotic scenes of thousands of foreigners gathered at the airport trying to leave or looking frantically for missing family members. Meanwhile, such basic necessities as water, electricity, and shelter were either unavailable or in short supply. According to a senior health official, Vichai Tian Thavorn, "The main problems [were] the spread of disease, sanitation, respiratory and skin diseases, [and] hygiene for people dealing with the bodies."[17]

In Sri Lanka, the official death toll more than tripled, from a low estimate of around five thousand up to 18,700 and then later still to 37,300 people killed or missing, plus 15,196 injured. Meanwhile, another 444,000 people were relocated to temporary refugee camps because their homes had been destroyed. The tsunami had devastated Sri Lanka's maritime and fishing industries, destroying an estimated 65 percent of Sri Lanka's fishing vessel fleet, some 29,700 vessels.[18]

In the north and east of Sri Lanka, the extent of the damage remained unclear, particularly in areas controlled by the Liberation Tigers of Tamil Eelam (LTTE). But in Trincomalee, an estimated thirty thousand people were homeless and there was no sign of refugee shelters promised by the government. In Batticaloa, a local member of parliament, S. Jeyanandamoorthy, told the Tamilnet website, "The newly built government hospital, public buildings, schools, LTTE political offices, homes, churches and temples have been completely destroyed by the sea."[19]

Extensive urban destruction, Aceh Province.

In India, the official death toll quickly increased to 11,500, more than half of them on the low-lying Andaman and Nicobar islands. According to the *Sydney Morning Herald,* at least seven thousand people had died there and as many as thirty thousand people were still unaccounted for. Later, a report described how on the Nicobar island of Chowdra a thousand of the 1,500 inhabitants had perished. Meanwhile, the death toll on Car Nicobar—a ten-square-mile atoll with a population of forty-five thousand—was thought to exceed ten thousand. On the Indian mainland, the southern state of Tamil Nadu experienced widespread damage. One day after the tsunamis, at least 3,720 bodies had been recovered. One fisherman told the press, "I was with my wife, my daughter, my father, my sister picking the small fish from my nets. Then the waters came. Now my family is gone."[20]

Reports from other countries in the area were more limited but sent the same message. At least sixty-five people were reported dead in Malaysia, ninety in neighboring Burma (Myanmar). In the tiny island nation of the Maldives the official death toll rose quickly to fifty-five. All the way across the Indian Ocean, at least 122 people were thought to have perished from the tsunamis in Somalia, Tanzania, and on the Seychelle Islands. One person had died as far away as South Africa. This second stage of international reports helped to clarify the extent of the disaster and the scope of the humanitarian disaster relief program that would be necessary for the survivors.

The International Community Finally Takes Note

It was late in the day on Monday, 27 December, then, that American newspapers and television stations began to report the true scope of the tsunami disaster. It was at this point—with U.S. Navy ships already on the move toward Southeast Asia—that President Bush authorized the Pentagon to initiate a humanitarian disaster relief operation.

New information now flowed into Washington and other world capitals but did not keep up with the rapid increases in casualty figures. By 29 December the estimated Southeast Asian death toll had jumped from 13,500 to the much larger, although still wildly inaccurate, estimate of sixty thousand.[21] This number was expected to rise even further, however, as rescue workers combed through the wreckage. The World Health Organization (WHO) warned that it could double, unless emergency supplies of food, water, and medicines reached the devastated areas. Throughout the region, relief workers warned that there were severe shortages of these necessities. David Nabarro, head of WHO's crisis operations, stated: "There is certainly a chance that we could have as many dying from communicable diseases as from the tsunami. . . . The immediate terror associated with the tsunami and the earthquake itself may be dwarfed by the longer-term suffering of the affected communities."[22]

In the three countries that were hardest hit and that would soon become the foci of American relief efforts—Indonesia, Sri Lanka, and Thailand—numerous corpses were still being washed ashore or uncovered as the waters receded. Transportation and communications were severely disrupted, especially in rural villages; aid workers could not travel to many areas; four days after the tsunami emergency teams had yet to reach many remote villages.

Due to the misleading early reports, major foreign donors, including the United States, the European Union, Australia, and Japan, were slow to react, initially pledging only U.S.$100 million in special relief aid. However, as the second stage of news reporting began to appear in the global media and the scope of the disaster became clearer, donations increased rapidly.

By then, however, Jan Egeland, the UN's emergency relief coordinator, had issued, the day after the tsunamis, a severe rebuke that for many helped clarify the true situation for these donor nations: "It is beyond me why we are so stingy. Really. Christmas time should remind many Western countries at least, how rich we have become. . . . There are several donors who are less generous than before in a growing world economy." Egeland estimated that the world community needed to donate at least $15 billion to deal with what he described as "the worst natural disaster in recent history."[23] Egeland later denied that he had specifically meant the United States. Nonetheless, in an attempt to blunt the perception that the U.S. government did not care, Secretary of State Colin Powell quickly rebutted the imputation that Washington had ignored the disaster: "The U.S. is not stingy," Powell declared. "We are the greatest contributor to international relief efforts in the world."[24]

Destroyed bridge, cutting north–south road traffic, Aceh Province.

It was in this setting that President Bush made his first public statement and pledged $15 million in aid. Unfortunately, by this time, the 29th, the magnitude of the disaster was far better understood within the region, and his offer was widely derided in the international media; one commentary from the French newspaper *Le Figaro* noted that it amounted to half the cost of a single F-16 fighter.[25]

Soon after Bush's announcement, the U.S. Agency for International Development (USAID) added another $20 million, bringing the American aid package to $35 million. Secretary Powell commented, "I think we've responded rather aggressively and appropriately. And the American people should be pleased and proud of the way we've done it."[26] Bush reiterated the point: "What you're beginning to see is a typical response from America. First of all, we provide immediate cash relief, to the tune of about $35 [million]. And then there will be an assessment of the damage. . . . I just got off the phone with the President of Sri Lanka, she asked for help to assess the damage. In other words, not only did they want immediate help, but they wanted help to assess damage so that we can better direct resources. And so our government is fully prepared to continue to provide assistance and help."[27]

The president now revealed for the first time that the U.S. military would be assisting the victims of the earthquake and tsunami: "We're dispatching a Marine expeditionary unit, the aircraft carrier *Abraham Lincoln,* and the maritime pre-position squadron from Guam to the area to help with relief efforts. . . . It takes money, by the way, to move an expeditionary force into the region. In other words, we're diverting assets, which is part of our overall aid package."[28]

The press generally ignored the president's offer of military assistance. The *New York Times* criticized the higher number of $35 million and the promise of long-term assistance as "a miserly drop in the bucket."[29] However quickly the Bush administration adjusted to new casualty numbers, it always seemed too little, too late. As one commentator later put it: "The United States may not have been 'stingy,' as a United Nations official charged, but the administration missed an opportunity to show the nation's generosity."[30] This put the U.S. government in the uncomfortable position of having to justify its extremely good humanitarian response. The early American relief response was "scaled up as the scale of the disaster became more widely known," Powell said on CNN's *Late Edition*. "We have nothing to be embarrassed about," he told NBC's *Meet the Press*.[31] However, as further news reports listed ever higher casualties, even $35 million soon appeared paltry; on 31 December Bush announced that the United States was pledging $350 million in relief assistance, a number that included dispatching the aircraft carrier *Abraham Lincoln*, the maritime prepositioning squadron at Guam, and an amphibious ship carrying a Marine expeditionary unit.[32] When these assets arrived in the theater, they focused their initial relief measures on emergency assistance, including providing much-needed fresh water, food, and medicine. According to Army colonel Gary Keck, a Pentagon public affairs officer, this contribution was not cheap, ultimately six million dollars per day. The price tag "cover[ed] a U.S. contingent of about 13,000 military personnel, of which about 12,000 are aboard 17 Navy ships in the region hit by the Dec. 26 earthquake and resulting tsunami." This made the American military's contribution to the humanitarian effort quite possibly the "largest in history."[33]

* * * * * * *

The lack of a regional tsunami warning system, coupled with widespread underestimations by local governments and the international press of the true casualty rates, resulted in what many have criticized as a slow U.S. response. One contributing factor was that local governments restricted media and aid access to some areas. Despite grave conditions in many parts of Aceh Province, for example, the Indonesian army lifted the ban on foreign aid workers only several days after the tsunami. Thereafter media reports from Banda Aceh described the local response to the catastrophe as limited, unplanned, and in some cases completely lacking.

After an initial donation of $15 million and then $35 million, by February 2005 USAID and the Defense Department would spend an estimated $346 million for immediate relief aid. As State Department spokesman J. Adam Ereli explained on 30 December 2004, "Moving a carrier strike force and a Marine expeditionary unit within 72 hours should not be considered dilly dallying. The planes are on the ground in Indonesia tonight."[34]

Throughout this period, Washington had to fight a largely uphill media battle to prove that it properly understood the scale of the disaster and was reacting adequately to it. Meanwhile, however, having perceived the true nature of the disaster, Admirals Fargo and Doran were responding quickly and massively, eventually deploying over two dozen warships and over a hundred aircraft.

Notes

1. Quote in John Aglionby, "Colin Powell Shocked by Scale of Destruction," *Guardian Unlimited*, 6 January 2005, available at www.guardian.co.uk/tsunami/story/0,15671,1384193,00.html.
2. "Tsunami Caught the Government Napping," *Times of India*, 26 December 2004, available at timesofindia.indiatimes.com/articleshow/971639.cms.
3. The depth of the Chagos Trench and the grade to shores off Diego Garcia do not allow for tsunamis to build up before passing the atoll; for that reason, the only effect of "the earthquake was seen as a tidal surge estimated at six feet." "Naval Base Is Spared Damage by Tsunamis," *International Herald Tribune*, 5 January 2005.
4. Lila Rajiva, "Tsunami Cover Up? NOAA and the Flood," *Dissident Voice*, 6 January 2005, available at www.dissidentvoice.org/Jan05/Rajiva0106.htm.
5. Richard Norton-Taylor, "U.S. Island Base Given Warning: Bulletins Sent to Diego Garcia 'Could Have Saved Lives,'" *London Guardian*, 7 January 2005.
6. Tim Palmer, "Aceh Death Toll Spirals after Tidal Waves," Australian Broadcasting Company, 27 December 2004, available at www.abc.net.au/news/newsitems/200412/s1272562.htm.
7. Reuters, "Beach Island Flattened," 27 December 2004, available at www.telegraphindia.com/.
8. "More Than 4,300 Killed in Sri Lanka's Worst Ever Disaster," *Terradaily*, 26 December 2004, www.terradaily.com/2004/.
9. "Kohl Survives Tsunami, Stays to Help Sri Lankans," *WWW Virtual Library*, 3 January 2005, www.lankalibrary.com/news/kohl.htm.
10. A. V. Ragunathan, "400 Feared Dead as Waves Lash Cuddalore Coast," *Hindu*, 27 December 2004, www.hinduonnet.com/.
11. "Beaches Turned into Mortuaries," *The Age*, 27 December 2004, www.theage.com.au/news/World/.
12. "Indian Death Toll from Tidal Waves Rises above 2,600," *Terradaily*, 26 December 2004, www.terradaily.com/2004/.
13. Peter Wallsten and Edwin Chen, "Bush Defends U.S. Response to Disaster," *Los Angeles Times*, 30 December 2004.
14. Admiral Walter Doran (Ret.), interview with Elleman, Arlington, Virginia, 23 October 2006.
15. Tim Johnson and Ron Hutcheson, "Warning Cloud over Relief Effort," *Philadelphia Inquirer*, 11 January 2005.
16. Robert Block, Greg Hitt, and Jess Bravin, "Aid for Asia Rises amid Daunting Needs, Logistics: Widespread Destruction Challenges Many Agencies Trying to Coordinate Relief," *Wall Street Journal*, 29 December 2004.
17. Connie Levett, "Out of Phuket, a Tour of Death," *Sydney Morning Herald*, 29 December 2004.
18. Lieutenant Colonel Daren K. Margolin, "Humanitarian Assistance/Disaster Relief Operations: Combined Support Group Operations in Sri Lanka," *Marine Corps Gazette* (October 2005).
19. "Landmine Fears Hampers Rescue Work in Batticaloa," *TamilNet*, 28 December 2004, www.tamilnet.com/.
20. Peter Foster, "Despair, Wreckage along a Fatal Shore," *The Age*, 29 December 2004, www.theage.com.au/news/World/.

21. "Officials Warn of Disease; Death Toll Tops 60,000," *USA Today,* 29 December 2004, available at www.usatoday.com/news/world/2004-12-28-tsunami_x.htm.

22. Tom Paulson, "Death Toll Could Double from Epidemic of Diseases," *Seattle Post-Intelligencer,* 29 December 2004, available at seattlepi.nwsource.com/national/.

23. "Thousands Die in Asian Tsunami," BBC News, 26 December 2004, available at news.bbc.co.uk/.

24. "Donations to Tsunami Relief 'Generous,' U.N. Says," CNN.com, 28 December 2004, edition.cnn.com/2004/WORLD/asiapcf/12/28/powell.aid/.

25. Stéphane Marchand, "Charité bien ordonnée," *Le Figaro,* 30 December 2004, available at www.lefigaro.com, as quoted by "U.S. New York Times Says U.S. Is 'Stingy,'" *AsiaMedia,* 30 December 2004, www.asiamedia.ucla.edu/.

26. Rafael D. Frankel, "U.S. Copters a Golden Vision—UN: Mercy Mission Saluted," *New York Daily News,* 3 January 2005.

27. "President Discusses Support for Earthquake and Tsunami Victims," 29 December 2004, available at www.whitehouse.gov/news/releases/2004/.

28. Ibid.

29. "U.S. New York Times Says U.S. Is 'Stingy.'"

30. David Ignatius, "More Water Bottles, Fewer Bullets," *Washington Post,* 4 January 2005.

31. William M. Welch, "Aid Reaches Remote Areas: Powell: U.S. Help Was Appropriate," *USA Today,* 3 January 2005.

32. "President Commits $350 Million for Tsunami Relief Effort," 31 December 2004, available at www.whitehouse.gov/news/releases/2004/.

33. "Pentagon Says It's Spending $6 Mln a Day in Tsunami Relief," *Wall Street Journal,* 6 January 2005.

34. Jim Vandehei and Robin Wright, "Bush Says America Will Lead Global Aid Effort," *Washington Post,* 30 December 2004.

CHAPTER THREE

Preparing the U.S. Navy Operation
The Value of Sea Basing

In one of the most complex relief operations ever attempted. . . . [n]aval forces are able to arrive with critical mass quickly, commence relief support immediately, and substance those operations indefinitely . . . by establishing a sea base as close to the operation as possible.

NAVY WARFARE DEVELOPMENT COMMAND, 2005

On 12 January 2005, only two weeks after the enormity of the tsunami crisis had become clear, almost fifteen thousand U.S. military personnel were providing humanitarian relief throughout the affected region.[1] These American servicemen and servicewomen were supported by twenty-five U.S. Navy ships and one Coast Guard cutter acting as temporary sea bases for forty-five fixed-wing aircraft and fifty-eight helicopters. By this point in the mission, the U.S. military had delivered 2.2 million pounds of relief supplies to the worst-hit nations, including sixteen thousand gallons of water, 113,000 pounds of food, and 140,500 pounds of other relief supplies during the previous twenty-four hours alone. Because of the extent of the local devastation, including the destruction of roads, bridges, and docks, sea basing was critical to the success of this humanitarian mission.

Several geographical and organizational factors proved critical, in turn, to the sea-basing effort: where the U.S. Navy assets had been when the tsunami disaster occurred, what exactly the Navy's orders were, when those orders were transmitted, and how quickly the ships responded. There was no significant delay between when President George Bush announced that he had authorized a relief operation and when the first U.S. Navy ships appeared offshore in the crisis area, because PACOM had sent movement orders to units involved two days before.

The U.S. Navy was particularly well positioned and equipped to react to this humanitarian crisis, it was ordered to respond as soon as the true magnitude of the tsunamis and the extent of the damage became clear, its ships got to sea almost immediately upon receiving orders, and they were authorized to use maximum safe speeds. Few surface ships then in existence could have reached the disaster area more quickly than they did.

Operation UNIFIED ASSISTANCE in Sri Lanka and Thailand

When news of the Sumatran earthquake and tsunamis first reached PACOM headquarters at Camp Smith in Hawaii, Admirals Fargo and Doran quickly decided to send all available ships to Southeast Asia.[2] But as discussed in the previous chapter, the local governments were not yet able to estimate the true numbers of casualties, whereas the international press was reporting only slightly more than ten thousand victims in all, which suggested that a small HA/DR operation might be sufficient. Conditions in Aceh Province were particularly unclear. Over the next three days, therefore, as estimates of the real extent of the crisis continued to rise, PACOM needed constantly to update and change the scope of what it would soon call Operation UNIFIED ASSISTANCE.

A liaison officer was requested from USAID, and "over the next 24 hours communications were established with foreign military commanders to assess their requirements."[3] Admiral Walter Doran also quickly called on a classmate, Admiral Arun Prakash, who had become the Indian navy's Chief of Naval Staff that July, and the two men agreed to an informal partnership.[4] Within a day or two of the disaster, these two old friends were able to discuss where the Indians would be operating and where the American forces could make the best contribution.[5]

On 28 December 2004, Admiral Fargo established Joint Task Force (JTF) 536, to be set up in Utapao, Thailand. Thailand was perfect, because the series of six-week-long U.S.-Thai joint/combined COBRA GOLD exercises had laid the foundation for emergency cooperation: "Our large multinational exercise that we conduct every year in Thailand," commented Admiral Fargo, "is specifically pointed toward humanitarian assistance, disaster relief and peacekeeping, and of course it brings a large number of the nations of the region together to work in this same manner. . . . So you can't point yourself toward a specific catastrophe like this, but you can put in place the basic training, the habitual relationships and, as I pointed out, the standard operating procedures that apply to a wide range of contingencies and crises."[6]

While the Marines constructed JTF headquarters from scratch, General Blackman established three combined support groups, in Medan, Indonesia; Galle, Sri Lanka; and Phuket, Thailand. Blackman assigned CSG-Indonesia to Brigadier General Christian B. Cowdrey, USMC, of the 3rd Marine Division; and CSG–Sri Lanka to Brigadier General Frank A. Panter, of the 3rd Force Service Support Group; he absorbed CSG-Thailand into his own III MEF staff.[7] On 3 January 2005, the increasing importance of coordinating these efforts with international partners was reflected in a redesignation of JTF 536 to Combined Support Force (CSF) 536.

Commander Combined Support Force Indonesia, Lieutenant General Robert R. Blackman, USMC, answering questions from the media, after speaking to sailors of the USS *Abraham Lincoln*.

Both PACOM's and CSF 536's "Operating Principles" for planning Operation UNIFIED ASSISTANCE (or OUA) were quite clear and concise: they included specific goals for tempo, leadership, unified action, coordination, priorities, and the desired end state. Pacific Command's mission statement declared: "USPACOM provides assistance to the governments of Indonesia, Sri Lanka, Thailand and other affected nations to mitigate the effects of the recent earthquake and tsunami in the Indian Ocean. Conduct of operations is in support of USG [the U.S. government] lead agency [USAID], and in coordination with international organizations, non-governmental organizations and partner nations." The CSF 536 mission statement was similar but further made clear the voluntary nature of the American presence: "On order, transition U.S. Military HA/DR activities to designated agencies and/or Host Nations, in order to facilitate continuity of relief and redeployment."[8]

To accomplish these tasks expeditiously, CSF 536 had to be prepared to work closely with both interagency relationships and with the many nongovernmental organizations (NGOs), international organizations, and other groups. These variously included, but were not limited to, UNICEF, WHO, Save the Children, Medicins Sans Frontières, and the International Federation of Red Cross and Red Crescent Societies, as well as the local and international press. The very complexity of these interagency and NGO relationships (see figure 1 for U.S. interagency relationships) made CSF 536's task an extremely sensitive and difficult one.

On 2 January 2005, Brigadier General Panter arrived in Colombo to establish Combined Support Groups–Sri Lanka. He immediately set up a civil-military operations center, called a combined disaster relief center, in order to facilitate the distribution of supplies and serve as the "central coordination point" with respect to other external organizations.[9] Two days later, an advance contingent of seventeen Marines landed at Colombo airport, supported by thirty Air Force communications and technical experts. Eventually Panter deployed 1,500 troops around Sri Lanka—mainly in the south, so as to avoid rebel strongholds in the north.[10]

FIGURE 1
U.S. Interagency Relationships

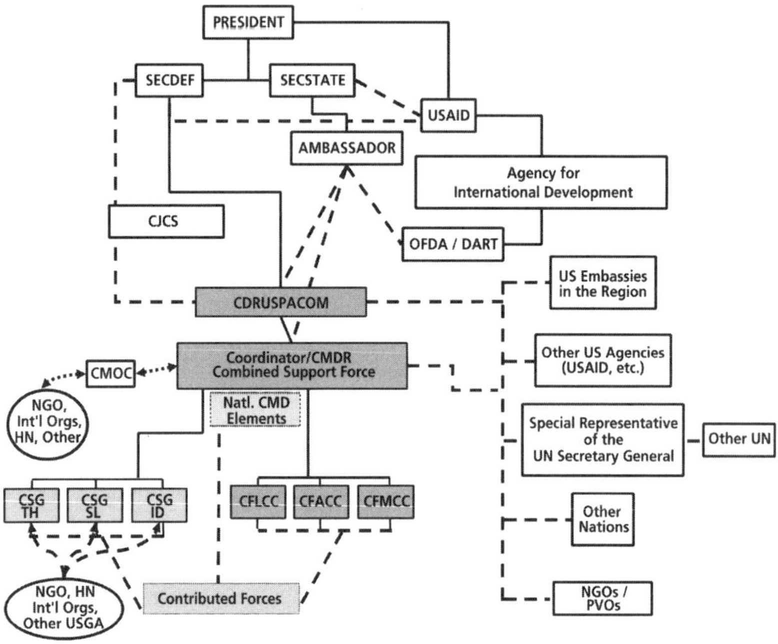

During the first week, they helped deliver approximately thirty-one thousand pounds of food, water, and medicine throughout southern Sri Lanka, mainly by HH-60 (Pave Hawk) helicopters from the 33rd Rescue Squadron, based in Kadena Air Force Base, on Okinawa.[11] In response to this influx of aid, the Sri Lankan foreign minister, Lakshman Kadirgamar, said: "I can say honestly and sincerely that the response has been overwhelming, beyond our expectations."[12]

In addition to providing aid, American Seabees and other engineers helped clear damage in Galle, one of Sri Lanka's seven "world heritage sites." They demolished two damaged buildings and moved 250 cubic yards of rubble at a women's college in Galle to clear space for future construction. These teams also cleared debris at several schools and at a Sri Lankan army base. "We've seen an immediate effect through our efforts," said LT Jorge Cuadros, the air detachment officer in charge of Navy Mobile Construction Battalion 7, from Guam: "Within two days of clearing the demolished schools, children were back to their classes learning."[13]

During the second week, the supplies delivered by U.S. forces increased to fifty thousand pounds as CH-46 Sea Knight helicopters from the 15th Marine Expeditionary Unit, sent from Iraq, flew a total of thirty-five missions. By week three, the 15th MEU had been replaced by Marine Corps and Coast Guard C-130s, and approximately forty thousand pounds of supplies were being flown into affected areas every day. By the end of January 2005, over six hundred thousand pounds of HA/DR supplies had been distributed—146,000 pounds of food, 124,000 pounds of medicine and other supplies, and 8,500 gallons of water.[14] By the middle of January, Lieutenant General Blackman could optimistically report that the Sri Lankan government was "meeting or exceeding the minimum needs of internally displaced persons."[15]

In Thailand, where the local government officials responded quickly to the disaster, specific requests for American assistance included a POW-MIA team from the Joint Prisoners of War/Missing in Action Accounting Command.[16] Meanwhile, the U.S. government deployed several eight-member forensic analysis teams, each team including civilian forensic anthropologists, dentists, mortuary affairs experts, and forensic photographers.[17] These groups recorded where victims' bodies were found, determined their identity, and organized proper disposal of the remains: "Mortuary Affairs are a special breed of soldiers," said Sergeant 1st Class Ronald E. Holliday; "We are trained more so than the average soldier to deal with the rigors of death."[18]

Meanwhile, at the air base at Utapao, U.S. Navy personnel were busy setting up the main logistics headquarters. According to Commander Russ Thompson, at the beginning there were only three buildings and a handful of people at the headquarters, but when he left on 6 January 2005 there were "over 900 to 1000 people at the center."[19] Not only did the airfield at Utapao serve as a major gateway for bringing humanitarian aid into the theater, but the CSG-Thailand staff in Phuket organized engineering teams to help with the cleanup. With American assistance, the most serious damage was repaired quickly, after which Thai military engineers continued to receive support from American engineering support teams under the auspices of the Joint U.S. Military Assistance Group.

By 15 January 2005, General Blackman had determined that U.S. assistance was no longer necessary in Sri Lanka and Thailand, since these countries were "beyond the stop-the-bleeding phase." Accordingly, he reported, "The conditions for transition from U.S. military to host-nation military control of operations in these countries have been met," and "within the next week or two" the U.S. military presence in these countries would "evolve into a more-normal theater-security support posture."[20] On 22 January CSF 536 concluded operations in Thailand, and a week later, on 28 January, it ended the humanitarian mission in Sri Lanka. However, the mission in northern Indonesia, the hardest-hit area of Southeast Asia, was far from over.

Operation UNIFIED ASSISTANCE in Indonesia

Because of the almost total destruction experienced in Aceh, sea basing was required. Admiral Fargo tapped naval assets that were already in the Pacific theater but were not part of operations in Afghanistan or Iraq, including USS *Abraham Lincoln,* visiting Hong Kong, and USS *Bonhomme Richard,* conducting exercises near Guam. "You're never going to have the whole picture right off the bat," Admiral Fargo later explained, so it is necessary to "posture for the worst and then ramp up capacity" (see table 1).[21]

The official CSG-Indonesia headquarters was in Medan, where Marines and naval personnel could coordinate directly with local authorities, including the governor of Aceh. But Rear Admiral Douglas Crowder headed the naval force in Indonesia in his capacity as the commander of Carrier Strike Group 9, the *Abraham Lincoln* group, which arrived off the coast of Banda Aceh on 1 January 2005. Because Crowder had the best situational awareness, he was for all practical purposes in charge of the operation in Indonesia.[22]

The breadth and complexity of PACOM's and CSF 536's goals, the principles under which UNIFIED ASSISTANCE was to operate, and the diversity of organizations involved made the sea-basing aspect of the operation particularly challenging. Communications discussing a possible humanitarian relief mission began to circulate the day after the Sumatran earthquake and tsunamis. On the night of 27 December Rear Admiral Crowder received a call from Pacific Command about the assignment. It was confirmed on the 28th that he and his carrier strike group would head toward Thailand rather than, as had been planned, to Korea. According to Crowder, this was a groundbreaking event, because it was a real "emergency surge operation rather than an exercise."[23]

During these first few days, the new assignment generated numerous additional tasks for the strike group staff and the flagship. Captain Raymond Ginetti, navigator of *Abraham Lincoln,* learned only after leaving Hong Kong that they were heading for Phuket, Thailand. Current charts for the waters off Thailand had to be broken out and corrected. Then, on 30 December, as more accurate information on the scale of the disaster became available, *Abraham Lincoln*'s destination changed again, this time to northern Indonesia, requiring further navigational preparations. A day later, as *Abraham Lincoln* steamed through the Malacca Strait and along the coast of Sumatra, the size of the catastrophe became vividly manifest: "bodies could be seen floating as far as 20 miles out to sea."[24] *Abraham Lincoln* arrived off the coast of northern Indonesia on 1 January.[25]

The bulk of the strike group learned of the mission long after their departure from Hong Kong, since for some time it was uncertain whether the operation would obtain Washington's approval.[26] One officer on board first heard about the tsunami disaster from his wife. At that point, however, most of the early news focused on Thailand; the

TABLE 1
List of U.S. Naval Units Active in Operation UNIFIED ASSISTANCE

CARRIER STRIKE GROUP (CSG)		
USS *Abraham Lincoln* (CVN 72) CSG	USS *Abraham Lincoln* (CVN 72) USS *Shiloh* (CG 67) USS *Benfold* (DDG 65) USS *Shoup* (DDG 86) USS *Louisville* (SSN 724) USNS *Rainier* (T-AOE 7)	Carrier Air Wing (CVW) 2: VFA-2 (F/A-18F) VFA-137 (F/A-18E) VFA-151 (F/A-18C) VFA-82 (F/A-18C) VAQ-131 (EA-6B) VAW-116 (E-2C) HSL-47 Det. (SH-60B) HS-2 (HH/SH-60H/F) HC-11 (MH-60S)
EXPEDITIONARY STRIKE GROUP (ESG)		
USS *Bonhomme Richard* (LHD 6) ESG	USS *Bonhomme Richard* (LHD 6) USS *Duluth* (LPD 6) USS *Rushmore* (LSD 47) USS *Milius* (DDG 69) USS *Bunker Hill* (CG 52) USS *Thach* (FFG 43) USCGC *Munro* (WHEC 724)	HMM-165
OTHER U.S. NAVY SHIPS IN THE AREA OF OPERATIONS		
USS *Swift* (HSV 2)	USNS *Watson* (T-AKR 310)	*WestPac Express*
MILITARY SEALIFT COMMAND SHIPS IN THE AREA OF OPERATIONS		
USNS *Mercy* (T-AH 19)	MV *1st Lt. Jack Lummus* (T-AK 3011)	SS *Maj. Stephen W. Pless* (T-AK 3007)
MV *Cpl. Louis J. Hauge Jr.* (T-AK 3000)	MV *Pfc. James Anderson Jr.* (T-AK 3002)	MV *1st Lt. Alex Bonnyman* (T-AK 3003)
USNS *1st Lt. Harry L. Martin* (T-AK 3015)	USNS *Tippecanoe* (T-AO 199)	USNS *John Ericsson* (T-AO 194)
USNS *San Jose* (T-AFS 7)	USNS *Mary Sears* (T-AGS 65)	USNS *John McDonnell* (T-AGS 51)
USNS *Concord* (T-AFS 5)	USNS *Niagara Falls* (T-AFS 3)	

tsunami's impact on Indonesia did not then seem significant.[27] Only two to three days later, after departing Hong Kong, the president and the Pentagon having given the operation a green light, did Rear Admiral Crowder inform the strike group about the goals and the extent of UNIFIED ASSISTANCE.[28]

The reaction of the U.S. Navy to the situation was as unconventional as the disasters that struck Southeast Asia had been unexpected. The USAID administrator, Andrew Natsios, commented, "In all my years in this line of work, I have never before seen a disaster that affects 12 countries, as this earthquake and tsunami have done. The velocity and force of the tsunamis that struck communities along the coast were ferocious. 150,000 or more people are dead. More than 1.5 million people are homeless or displaced."[29] However, as soon as the *Abraham Lincoln* group left Hong Kong the complicated planning process began, with consideration of how best to assist countries affected by the earthquake and tsunamis. Only after the force was formally ordered to assist, of course, could detailed planning for the actual relief operation begin.

Setting Up the Joint Planning Group

Because local information on the disaster was so unreliable, it was difficult to know in advance exactly what conditions would be faced. In the meantime, one of the first things Rear Admiral Crowder did was set up what was known as the Joint Planning Group (JPG) to coordinate all the elements that make up a humanitarian relief mission. According to Captain Kevin Campbell, Admiral Crowder's Assistant Chief of Staff for Plans, the Joint Planning Group met for the first time on 28 December 2004. Twice a day, until the ships reached their destination, the JPG met to envision the various contingencies involved in a humanitarian relief mission. Because guidance from PACOM was incomplete—the ships had merely been ordered to the general area to assist—the JPG had to try to take into account the many possibilities that might arise: "Plans and thoughts included helicopter operations, amphibious ships, everything that could happen, including what and where."[30]

Lieutenant General Blackman later described this process as "planning your family vacation while you are packing the car"—that is, simultaneously "planning, executing and deploying."[31] For the planners in Utapao, this included making contact with supporting units in Singapore, flying advance teams to the mission area to decide how to get supplies from the carrier to the beach, and determining how to fly C-130s into the worst-hit areas near Banda Aceh airport.[32] Within a few days an Air Force team certified this airfield for C-17 cargo planes, which could carry almost five times the load of a C-130.[33] On 11 January 2005, the first C-17 landed at Banda Aceh carrying over eighteen thousand pounds of rice, twenty-two thousand pounds of water, a pickup truck, and floodlights for the airport.[34]

Meanwhile, CDR Vince Quidachay, director of logistics for the carrier strike group, was concerned that the force's logistical train was still heading for Korea. The flagship sent immediate requests for oilers to refuel the strike group. In effect, the JPG had to reorganize almost overnight a logistics schedule that is normally planned three or four months in advance.[35]

One result of the JPG's insights was that Joint Task Force 536 was changed to Combined Support Forces 536 (CSF 536), in order to allow the naval force to coordinate directly with all of the regional liaison officers (LNOs). Commander, Carrier Strike Group 9 would also direct all of the major units in the immediate area, including Air Force elements operating in support, and naval forces in Indonesia, as well as Commander, Task Force 76 (CTF 76), at White Beach Naval Facility, Okinawa, Japan. This was certainly "not typical operations for [a] carrier flight group."[36]

Creating Crisis Action Teams

Another of the Joint Planning Group's first decisions was to create crisis action teams (CATs) to organize specific aspects of the mission. The CATs assumed responsibility for many day-to-day activities, including finding qualified personnel to join shore parties that would be sent into the disaster zones, organizing the teams' daily work, and keeping statistics. When a helicopter from *Abraham Lincoln* crash-landed at Banda Aceh airport, a CAT organized the cleanup and repairs.

The shore parties were staffed by volunteers. The CATs did not take just anybody for this duty but sought people with relevant experience, such as responding to other natural disasters, like hurricanes or tornadoes. Very quickly, between 1,300 and 1,400 crew members volunteered.[37] Many were eager to help rebuild Banda Aceh, "including plumbers, welders, electricians and other technicians who could help repair infrastructure."[38]

Once volunteers had been assessed and those with the required background selected, doctors prepared them to go ashore. All of the volunteers had to receive new immunizations, including malaria boosters.[39] There was also concern that sunscreen and mosquito repellent would run short; the CATs had to locate additional supplies. Finally, every day about two hundred volunteers participated in classes and screening. Training in the Advance Trauma Lifesupport System and in field conditions was provided during the transit for all shore party volunteers.[40] Also during this period helicopter pilots and crews went through their own refresher training courses on cargo handling, including first aid and CPR refreshers. Meanwhile, maintenance crews modified helicopters to carry emergency supplies rather than people. This required removing most of the passenger seats to make room.[41]

A final element was cultural training. Indonesia is the world's largest Islamic nation; about 88 percent of its 238 million people are Muslims.[42] During the transit, the ship's chaplain assembled a slide show on what the shore parties might expect, along with a "culture brief" on Banda Aceh.[43] Cultural training included possible friction points, and a Muslim sailor on board talked to the volunteers on "how to behave so not to appear as [an] ugly American."[44] To reduce the likelihood of problems with locals, some American cultural habits had to change. One female commander assigned to the USNS *Mercy* later recalled, "One adjustment was to not make eye contact with men as women are not considered equals in Banda Aceh. At first, there was an adjustment period for them to get comfortable with us and for us to get comfortable with them."[45]

According to Chaplain Tom Walcott, the air wing chaplain, the CATs also would have to monitor stress levels arising from working in an area that had experienced widespread death and destruction. The helicopter crews in particular were to see numerous dead and injured.[46] Lieutenant General Blackman ordered that mental health counseling be offered to all service members, and that it be provided "earlier, rather than later."[47] Before leaving Thailand after the contingency, for example, most military personnel met with a chaplain to discuss their experiences. In Hawaii, Captain James Danner, the Pacific Fleet and Pacific Command chaplain, continued to provide counseling after the operation ended.[48]

Eighty-eight percent of Indonesia's population are Muslims.

The Joint Planning Group's decision to set up crisis action teams took the best possible advantage of the transit period: locating suitable volunteers for the shore parties, training them, giving them all full medical checkups and a cultural awareness course, and making provision for possible side effects. Later, when relief operation missions began in Indonesia, the CAT cells were seen "as the admiral's command center of sorts" and were the "main POC [point of contact] for those calling in to Strike Group," including LNOs, government officials, and others.[49]

Force Protection and Cultural Fears

Throughout the humanitarian mission, U.S. forces dealt with force protection on an ongoing basis.[50] In Sri Lanka, for example, Marines were careful to stay out of the north and east, so as to avoid the Liberation Tigers of Tamil Eelam; Indian forces were assigned to work in areas controlled by the guerrillas.[51] In Indonesia, to avoid similar problems with the Free Aceh Movement (the Gerakan Aceh Merdeka, or GAM), the U.S. government promised that troops would not set up base camps on Indonesian territory but would remain sea based.[52] The need for this assurance reflected, in part, deep cultural fears among Indonesians, in particular that Christians within the American military would try to convert them, or that the force was actually planning to stage an invasion, as Iraq had been invaded the year before. All of these sensitivities were assuaged by keeping the majority of U.S. forces offshore on board ship.

Nonetheless, it was the ongoing insurgency in Aceh that made force protection a top priority. A cease-fire had been declared on 27 December, but during early January there were reports of open fighting between government forces and insurgents from the Free Aceh Movement, blocking a relief convoy for eight hours.[53] Indonesia's state-run news agency on 21 January quoted the army's chief of staff, Ryamizard Ryacudu, as saying that the army had killed at least 120 rebels in the previous two weeks.[54] Force protection was also made potentially troublesome by the large numbers of refugees. As one American military civil-affairs officer acknowledged, "Restless refugee-camp males are the prime breeding ground of some of these insurgencies."[55]

In early January 2005, it was reported that members of the Laskar Mujahidin organization had been sent to Aceh Province in the wake of the disaster. This group had been founded during the 1990s and was at one point headed by Abu Bakar Bashir, an Islamic cleric who was in December 2004 on trial as a leader of Jemaah Islamiyah, which had suspected links to al-Qa'ida. Laskar Mujahidin's greatest concern was that the foreigners in Aceh would attempt to convert Muslims to Christianity. Jundi, one of the Laskar Mujahidin members in Banda Aceh, declared, however, that his group would not interfere if UNIFIED ASSISTANCE remained strictly a humanitarian operation: "We are here to help our Muslim brothers. As long as they are here to help, we will have no problem with them."[56]

Amien Rais, the head of Indonesia's legislative body, took the same view: "If the United States and Australia act beyond their humanitarian task, then we have to resist. But if otherwise, we ought to thank them."[57] To help reassure the Indonesians that conversion was not part of their mission, when the hospital ship USNS *Mercy* deployed in early January 2005, it had a Muslim naval chaplain onboard: "We demonstrated that as Americans we value religious freedom."[58]

Fear of conversion attempts was also a major concern for many ordinary Muslims in Indonesia. Dien Syamsuddin, secretary general of the Indonesian Council of Ulemas, or religious scholars, warned all "nongovernmental organizations, either domestic or international, with hidden agendas coming here with humanitarian purposes but instead proselytizing, this is what we do not like." He condemned reports that the U.S.-based WorldHelp had "planned to adopt 300 Acehnese children orphaned by the disaster and raise them in a Christian children's home." This group immediately dropped its proposal when local Muslim groups opposed it.[59]

For its part, the Indonesian military had its own concerns about plans to land Marines from *Bonhomme Richard.* Positioned off the city of Meulaboh, where only several thousand residents had survived out of an original population of sixty thousand, this ship had landing craft ready to put about a thousand Marines ashore. This movement was delayed, however, because it might appear to be an invasion. Aceh Province had been under the control of the Indonesian military, and it was thought that televised images of U.S. landing craft heading for the Acehnese coast "could touch a raw nerve with the proud and suspicious Indonesian military."[60]

Finally, on 10 January 2005, a U.S. Navy LCAC—air-cushion landing craft—went ashore with thirty pallets of food and water. Only a few dozen personnel on *Bonhomme Richard* were allowed to go ashore each day. Also, instead of driving vehicles themselves to deliver aid—and risking traffic accidents that might spark anti-American anger, as had happened in places like South Korea—the Marines left final distribution of the supplies mainly to the Indonesian military.[61]

Some Indonesians were suspicious of American surveying and reconnaissance missions. Captain B. Junair, an Indonesian helicopter pilot, was skeptical about the true motives behind these missions, in which P-3 Orion reconnaissance planes flew from bases in Thailand to photograph large swaths of Aceh. American officers explained that the aircraft were mapping the destruction of roads and bridges, and using infrared sensors to locate refugees in the mountains. But Captain Junair was suspicious that such information could be equally useful during a military invasion: "That will be helpful for them in the future."[62]

It would not be an understatement to say that many Indonesians were nervous at the Americans' arrival. Ariso Agus, a thirty-six-year-old construction worker, said, "We have some concerns. I don't agree with the American occupation of Iraq. And I don't agree with American morals, like men and women sleeping together in the same bed. We don't want any of that. . . . If the Americans are coming, they shouldn't stay long."[63] However, Zulbahri, thirty-eight, a carpenter in Banda Aceh who had lost his wife and three sons, appreciated help from abroad: "We are here in difficulties. If the U.S. comes, it is good. But please don't bring any other interests like religion and politics."[64]

* * * * * *

Given the ever-present force protection and cultural concerns, the U.S. Navy's ability to remain offshore of Aceh on sea bases decreased the American footprint, reduced friction, and so greatly facilitated achieving the mission's objectives. Sea basing helped eliminate unwanted accidents or incidents, even as U.S. Navy assets off Indonesia relayed ashore over four hundred thousand gallons of fresh water, over ten million pounds of food and supplies, and the treatment of thousands of patients.[65] This sea-based operation proved to be especially successful in fostering unparalleled cooperation between Christians and Muslims.

While American military forces remained at sea, American civilians worked on land. To support the sea-based capabilities provided by *Abraham Lincoln* and its strike group, USAID mobilized its civilian staff of 150, including members of the Office of U.S. Foreign Disaster Assistance (OFDA), to cooperate with the U.S. military. Army colonel Doug Wallace, a civil affairs officer assigned to the Combined Support Force 536, explained that while the mainly civilian OFDA specialists had the local expertise and so could handle the land component, the military could offer "depth of staff, and planning, logistical and communications capabilities" at sea, which is why a "close working relationship between OFDA and the military is typical during humanitarian relief operations."[66]

This teaming of civilian experts with the military made possible the rapid delivery of life-and-death aid to Aceh Province. It was arguably the first practical use for civilian purposes of a strategic concept—sea basing—that had originally been created for military purposes. U.S. Navy assets off Indonesia supported the delivery of fresh water, food, and medicine. In early February 2005, when the USS *Abraham Lincoln* and *Bonhomme Richard* redeployed, they were replaced by the USNS *Mercy*, supported by helicopters working off USS *Essex* (LHD 2) and LCACs from the dock landing ship *Fort McHenry* (LSD 43). As this operation proved, sea basing's greatest single advantage lay in coordinating throughout the operation the enormous logistical flow into the affected areas.

Notes

1. Epigraph from "Humanitarian Assistance/ Disaster Relief (HA/DR) Operations Planning," TACMEMO 3-07.6-05 (Newport, R.I.: Navy Warfare Development Command, August 2005), annex A.
2. Admiral Walter Doran (Ret.), interview with Elleman, Arlington, Virginia, 23 October 2006.
3. "Tsunami Relief," briefing notes provided by Admiral Walter Doran (Ret.), 23 October 2006.
4. Admiral Vernon Clark (Ret.), telephone interview with Elleman, 17 October 2006.
5. Doran interview.
6. Thom Shanker and James Brooke, "Tsunami Tests U.S. Forces' Logistics, but Gives Pentagon a Chance to Show a Human Face," *New York Times*, 9 January 2005.
7. Lieutenant Colonel Daren K. Margolin, "Humanitarian Assistance/Disaster Relief Operations: Combined Support Group Operations in Sri Lanka," *Marine Corps Gazette* (October 2005).
8. Ivan Luke, "Operation UNIFIED ASSISTANCE Case Study," NWC 3096 (Newport, R.I.: Naval War College, n.d.), p. A-3.
9. Margolin, "Humanitarian Assistance/ Disaster Relief Operations." So as to include as many coalition partners as possible, the CDRC structure was considered preferable to a civil military operations center (CMOC).

10. Lennox Samuels, "Marines Arrive in Sri Lanka," *Dallas Morning News,* 6 January 2005.

11. Master Sergeant Orville F. Desjarlais, Jr., "Mission of Mercy: Air Force Rescue," *Airman* (April 2005).

12. Mimi Hall and Barbara Slavin, "Logistics Delaying Supplies in S. Asia: U.S. Doubles Its Copters for Relief," *USA Today,* 5 January 2005.

13. Juliana Gittler, "U.S. Military Begins to Leave Tsunami-Stricken Region," *Pacific Stars and Stripes,* 24 January 2005.

14. Ibid.

15. Kathleen T. Rhem, "'Conditions for Transition' Met in Thailand, Sri Lanka," American Forces Press Service, 15 January 2005.

16. Toby Harnden, "'Time Is the Big Enemy' as Scientists Check the DNA of the Dead," *London Sunday Telegraph,* 9 January 2005.

17. Guy Taylor, "Forensic Teams Sent to Tsunami-Hit Areas," *Washington Times,* 6 January 2005.

18. Jeffrey Kelley, "Mortuary Affairs Unit Heads to South Asia from Fort Lee," *Richmond Times-Dispatch,* 4 January 2005.

19. Naval Historical Center [hereafter NHC], Oral Interview Summary, NCDD 206, conducted by Captains William Engvall, Kevin Gillis, John Lackie, Mike McDaniel, Woodrow Shields, and Commanders Clint Ashley and Charlie Beckum, compiled by Historical Products Officer, Commander Doreen Fussman and assistant Yeoman First Class Paul Coleman. Commander Russ Thompson, executive officer, HS-2, NHC Oral History, 20 February 2005.

20. Rhem, "'Conditions for Transition' Met in Thailand, Sri Lanka."

21. Admiral Thomas Fargo, telephone interview with Elleman, 30 July 2006.

22. Doran interview.

23. Rear Admiral Douglas Crowder, CCSG-9, NHC Oral History, 20 February 2005.

24. Wayne Arnold, "In Areas All but Unreachable, Helicopter Crews Bring Help," *New York Times,* 5 January 2005.

25. Captain Raymond Ginetti, navigator, USS *Abraham Lincoln,* NHC Oral History, 20 February 2005.

26. Doran interview.

27. Captain Bob Aronson, CCSG-9 staff N6/ ACOS C41, NHC Oral History, 15 February 2005.

28. HS-2, Group 2, NHC Oral History, 20 February 2005.

29. Andrew S. Natsios, "Post-Tsunami Reconstruction," U.S. Chamber of Commerce Forum, 18 January 2005, available at gopher.info.usaid.gov/press/speeches/2005/sp050118.html.

30. Captain Kevin Campbell, CCSG-9 staff N5, NHC Oral History, 20 February 2005.

31. Blackman quoted in Arnold, "In Areas All but Unreachable, Helicopter Crews Bring Help." "Planning, executing and deploying" in James Brooke, "Ad-hoc Planning Goes into Marine Aid Effort," *New York Times,* 6 January 2005.

32. Campbell oral history.

33. Greg Jaffe, "U.S. Armed Forces Stretched Further by Aid to Asia," *Wall Street Journal,* 6 January 2005.

34. Rafael D. Frankel, "U.S. Cargo Plane Brings Lifesaving Supplies to Region," *New York Daily News,* 11 January 2005.

35. Commander Vince Quidachay, CCSG-9 staff N4, NHC Oral History, 20 February 2005.

36. Campbell oral history.

37. Commander Chris Meyer, chief engineer, USS *Abraham Lincoln,* NHC Oral History, 20 February 2005.

38. Richard C. Paddock, "Navy Airlifts Supplies to Ravaged Coast," *Los Angeles Times,* 3 January 2005.

39. Commander Michael Horan, commanding officer, HS-2, NHC Oral History, 20 February 2005.

40. Medical Group 1, USS *Abraham Lincoln,* NHC Oral History, 20 February 2005.

41. HSL-47, Group 2, NHC Oral History, 20 February 2005.

42. Alan Sipress and Glenn Kessler, "A Clear Line between Life and Death," Washington Post Foreign Service, 5 January 2005.

43. Meyer oral history.

44. Commander Dan Boyles, executive officer, HSL-47, NHC Oral History, 20 February 2005.

45. Commander Karen McDonald, USNS *Mercy* (T-AH 19), NHC Oral History, 6 June 2005.

46. Lieutenant Commander Tom Walcott, chaplain, CVW-2, NHC Oral History, 20 February 2005.
47. Lisa Burgess, "Commanders Meeting to Formulate Long-Range Goals in Asian Relief Effort," *European Stars and Stripes,* 8 January 2005.
48. Craig Gima, "Returning Military Gets Counseling First," *Honolulu Star-Bulletin,* 21 January 2005.
49. Lieutenant Sean Clevenger, CAT Cell member, CSG-9, NHC Oral History, 22 February 2005.
50. Captain David Llewellyn, USNS *Mercy* (T-AH 19), NHC Oral History, 2 June 2005.
51. Josh White, "Wolfowitz Cites Sri Lanka's Progress on Reconstruction: Regional Relief Efforts Called 'Platform' for Future," *Washington Post,* 18 January 2005.
52. Denis D. Gray, "Indonesia Denies U.S. Pilots Use of Airspace; Carrier Leaves," *Washington Times,* 13 January 2005.
53. Tim Johnson and Ron Hutcheson, "Warning Cloud over Relief Effort," *Philadelphia Inquirer,* 11 January 2005.
54. "Aid Groups Pledge to Bear Greater Share of Burden as Troops Begin Withdrawal," *Baltimore Sun,* 21 January 2005.
55. Guy Taylor, "Tsunami Relief Runs into Snags," *Washington Times,* 9 January 2005.
56. Chris Brummitt, "Group Linked to Al Qaeda Starts 'Relief' Work in Aceh," *Washington Times,* 7 January 2005.
57. Chris Brummitt, "Security Threats Won't Limit Efforts," *Boston Globe,* 8 January 2005.
58. Lieutenant Commander John Owen, USNS *Mercy* (T-AH 19), NHC Oral History, 4 June 2005.
59. Alan Cooperman, "Tsunami Orphans Won't Be Sent to Christian Home," *Washington Post,* 14 January 2005.
60. David Ibison and Jake Lloyd-Smith, "Indonesia Delays U.S. Navy Aid Mission," *London Financial Times,* 10 January 2005.
61. Richard Spencer, "U.S. Marines Stage Beach Landings to Bring in Supplies," *London Daily Telegraph,* 11 January 2005.
62. Alan Sipress and Noor Huda Ismail, "Relief Transcends U.S.-Indonesia Divide: Rights Concerns Sidestepped as Militaries Respond Jointly to Disaster," *Washington Post,* 4 January 2005.
63. Barbara Demick, "U.S. Troops Are Relieved to Be Greeted with Smiles," *Los Angeles Times,* 11 January 2005.
64. Richard C. Paddock, "Navy Airlifts Supplies to Ravaged Coast," *Los Angeles Times,* 3 January 2005.
65. Gittler, "U.S. Military Begins to Leave Tsunami-Stricken Region."
66. Fred Zimmerman, "USAID Teams Help Military Deliver Relief to Where It's Needed in Disaster Area," *European Stars and Stripes,* 8 January 2005.

CHAPTER FOUR

Replenishment and Sustainability Issues

Logistics as Leverage

The U.S. Military there has been the backbone of the logistical operations providing assistance to all afflicted after the disaster.

JUWONO SUDARSONO, MINISTER OF DEFENSE,
REPUBLIC OF INDONESIA, 16 JANUARY 2005

Sea bases provided enormous logistical capabilities, the power and utility of which immediately became apparent in Aceh Province, where much of the road system had become impassable.[1] A study of Operation SEA ANGEL in Bangladesh during May 1991 referred to the U.S. Navy's supply system as "logistical leverage."[2] In Indonesia, sea-based U.S. Navy ships leveraged their capabilities as logistical hubs, providing substantial supplies from their own stores as well as coordinating delivery by specialized supply ships and cargo planes directly to Banda Aceh airport and to other temporary points of distribution.

Sea basing's logistical leverage was magnified by the fact that tsunamis followed on the heels of the earthquake. This meant that while the low-lying areas were hit with waves, many mountainous regions had already suffered earthquake damage. In some areas of Aceh Province, the earthquake destroyed over 70 percent of the bridges and collapsed large sections of roadway. Such widespread damage posed a major logistical challenge to relief efforts and hampered the entire region's economic recovery.

Logistical supplies destined for Banda Aceh came from a variety of places, including U.S. Navy ships, vessels of U.S. Transportation Command's Military Sealift Command (MSC), Air Force C-130s, C-5s, and C-17s from Utapao and Singapore; USAID; dozens of countries; NGOs; and the Indonesian government. No matter where supplies came from, however, it was generally the sea-based U.S. Navy ships that became responsible for coordinating the logistical flow.

Onboard Stores

One of the best early sources of supplies was the ships themselves. As soon as the USS *Abraham Lincoln* arrived off northern Indonesia, sailors stacked fresh water, protein drinks, and bags of rice on helicopters to be flown to shore.[3] The carrier itself produced in its distilling plants up to eight hundred five-gallon jugs of fresh water per hour.[4] To maximize the amount of fresh water available to tsunami victims, sailors cut their water usage, reportedly even forgoing showers.[5]

When it came to logistics, "flexibility was the key to supporting the needs of the mission." *Abraham Lincoln* spent well over $150,000 from its operational funds to renew its rapidly depleting stores. For example, the ship soon ran low on bottled water and fresh food, as well as fresh linens, pillows, and cots, all of which were needed to house official guests, representatives from NGOs, and the press. Also, every day throughout most of January 2005 the galley staff had to make over five hundred box lunches for the shore parties alone.[6]

Meanwhile, when the USS *Bonhomme Richard* stopped in Guam on its way to Indonesia, its supply personnel purchased necessities at a local hardware store: "The mission began when Marines went into Ace Hardware—an American do-it-yourself chain—near the U.S. naval base on the Pacific island of Guam and bought more than £50,000 worth of timber, plastic sheeting and other supplies on a credit card."[7] Within half a day, the supplies had been trucked back to the ship and loaded aboard.

Onboard medical supplies, however, were not specifically intended for a natural disaster. When in early January the USNS *Mercy* was quickly loaded in San Diego with three hundred pallets of supplies, it all had to be inventoried, pallet by pallet, so the supply

Water staged for onload to helicopters on *Abraham Lincoln*.

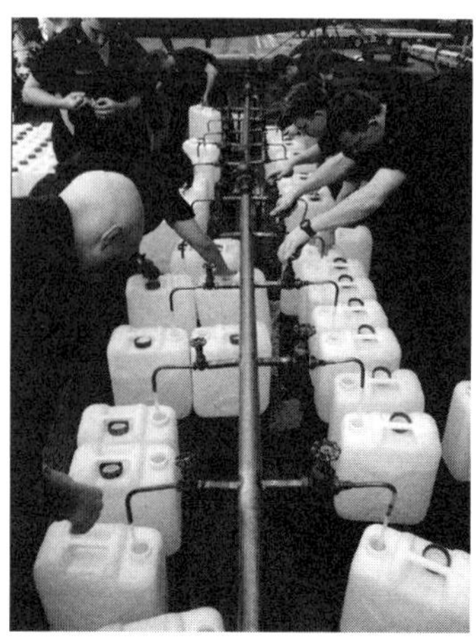

Water manifold on *Abraham Lincoln*.

clerks could check with the Medical Department to determine what was missing. However, since nobody knew exactly what the mission would entail, even the Medical Department did not know what it might need to accomplish the as "yet to be determined mission."[8]

A second medical problem was the short shelf life of blood, plasma, and immunizations. Such medicines had to be purchased, shipped, and stored. Timing when to order those types of medical supplies was critical, however, since they might otherwise be wasted. The type and length of the impending HA/DR mission were unknown, so the quantity of different medicines that might be needed could only be guessed at.[9]

Finally, there was the problem of having the right supplies on hand at the right time. The biggest challenge was that *Mercy*'s normal medical inventory was intended mainly for combat scenarios; there were no, for example, pediatric or geriatric medicines. Since *Mercy*'s pharmaceutical supplies were primarily tailored for trauma, the ship had to order specialized HA/DR pharmaceuticals, including strong antibiotics for treatment of "tsunami lung," an aspiration pneumonia due to bacterial or fungal infection.[10] Such special-purpose medical supplies were ordered and stocked on the basis of the best information available, and later additional supplies were loaded at Guam.

Over the course of its mission in Indonesia, *Mercy* received and processed over 1,800 additional pallets of supplies.[11] Although it was a challenge to switch so quickly from trauma care to a humanitarian mission, in the end the ship adapted and utilized its own medical supply inventory while effectively ordering, receiving, and consuming enormous quantities of additional medical supplies during the deployment.[12]

The Military Sealift Command Lends a Hand

The U.S. Transportation Command quickly authorized the MSC to send ships to the scene of the disaster. According to Brigadier General Paul J. Selva, TransCom's operations director, it was the "biggest natural disaster response in the command's history."[13] Maritime Prepositioning Ships Squadron 3 provided six of its thirteen ships. MV *1st Lt. Jack Lummus* (T-AK 3011), SS *Maj. Stephen W. Pless* (T-AK 3007), MV *Cpl. Louis J. Hauge Jr.* (T-AK 3000), MV *Pfc. James Anderson Jr.* (T-AK 3003), and MV *1st Lt. Alex Bonnyman* (T-AK 3003) sailed from Guam, while USNS *1st Lt. Harry L. Martin* (T-AK 3015) sailed from Korea.[14]

To provide food and fuel to the sea-based ships, the MSC also sent two oilers, USNS *Tippecanoe* (T-AO 199) and *John Ericsson* (T-AO 194); one combat stores ship, USNS *San Jose* (T-AFS 7); and one fast combat support ship, USNS *Rainier* (T-AOE 7). Later, the large roll-on sealift ship USNS *Watson* (T-AKR 310) deployed to Sri Lanka to provide support, while *WestPac Express*, a high-speed vessel, departed Okinawa under charter on 5 January 2005, carrying supplies to Thailand; it was designated as the "III MEF standby delivery vessel."[15] Finally, *HSV-2 Swift*, a second high-speed chartered vessel, left Ingleside, Texas, on 3 January, arriving in Southeast Asia late that month.[16]

For the next three months, MSC ships like *San Jose* acted as a "sea base for HA/DR material and a key logistical link" for the *Abraham Lincoln* carrier strike group, as well as the *Bonhomme Richard* expeditionary strike group, and replenished at sea the French frigate *Georges Leygues* (D640) and the Australian amphibious ship HMAS *Kanimbla* (L-51). *San Jose* was also the last T-AFS on station at the end of the operation, providing helicopter and supply support to *Mercy*.[17]

During the next two months, *San Jose* delivered over two thousand pallets of provisions, mail, repair parts, and other essential supplies to ships that were involved in this operation. Another seven hundred pallets of HA/DR materials, including food, water, and medicines donated by Project Hand-Clasp, were cross-decked to other ships to be transferred ashore or were taken directly ashore by *San Jose*'s two MH-60 helicopters.

Additionally, over two hundred helicopter sorties were flown from *San Jose* to Banda Aceh and other hard-hit coastal areas, transporting hundreds of passengers, including aid workers, *Mercy* medical teams, and patients. Commander Tae Lee was to recall that "although most of us did not get to see the people we were helping, we were nevertheless highly motivated because we realized that each HA/DR pallet coming off our deck provided a lifeline to those who were in desperate need . . . and that in itself is reward enough."[18]

Lastly, two MSC oceanographic ships were initially ordered to conduct hydrographic surveys of the ocean bottom where the 9.0-plus-magnitude earthquake had occurred.

American and Indonesian armed forces personnel transferring supplies to a waiting U.S. Navy helicopter.

USNS *Mary Sears* (T-AGS 65) and *John McDonnell* (T-AGS 51) sailed from Sasebo, Japan, for that purpose. When *Mary Sears* returned to regular duties, *John McDonnell* remained and continued to survey near the affected area.

Thailand and Singapore as Logistical Hubs

In addition to supplies delivered by ship, many early supplies delivered to Banda Aceh came by air, via the airfield Utapao, where work parties set up in a soccer field adjacent to the main airstrip.[19] For supplies delivered by sea, Singapore played an especially important role, processing millions of pounds of materials. The U.S. government's strong relations with Thailand and Singapore proved critical to the success of the mission.

In Utapao, there were initially only a couple of old buildings to work from. However, as described by Lieutenant Commander Scott Murdock, it was "amazing to watch Marines build up Camp Red Horse, from Vietnam area when Navy [*sic*] had a base there." On 1 January 2005 there were less than a hundred people on the base, but for the next month it never stopped growing. Eventually, Camp Red Horse had its own SIPR (classified Internet) "café," complete video-teleconference facilities, and other necessary communications equipment.[20]

Much of the aid shipped by water originated in Singapore. The Logistics Group Western Pacific and the Naval Regional Contracting Center Singapore were in charge of moving HA/DR supplies to sea-based ships. Rear Admiral Kevin Quinn, Commander, Logistics Group Western Pacific, would recall: "My goal was to rapidly establish a logistics capability that allows us to process much larger volumes of cargo than we normally do." During the first two weeks of the operation, about two million pounds of cargo were processed in Singapore alone.[21]

As the main provider of Navy logistics throughout Southeast Asia, Singapore already had a robust at-sea logistics capability, augmented by logistics and operations experts and assisted by specialists from the MSC and the Naval Oceanographic Office. Because their focus was resupplying U.S. Navy ships at sea, humanitarian assistance intended for tsunami victims could most quickly flow from Singapore to ships and then be flown by helicopter to Indonesia. During the first days of the crisis the Logistics Group Western Pacific began to analyze ways to move supplies directly to such countries as Indonesia and Sri Lanka, as well as by means of the *Abraham Lincoln* and *Bonhomme Richard* groups. Within days, representatives from all the military commands in Singapore, as well as experts from a variety of agencies, including the Defense Logistics Agency's Emergency Supply Operations Center, Naval Expeditionary Logistics Support Force, and the Fleet and Industrial Supply Center, began meeting to coordinate their logistical needs.

One early decision was that the Naval Regional Contracting Center would begin purchasing in Singapore more than $250,000 in relief supplies, including bottled water, food, and medicine. Many of these supplies were loaded onto MSC supply ships, passed on to the *Abraham Lincoln* strike group, and then delivered by helicopter directly to the worst disaster areas. All of these supplies were delivered in conjunction with, and in addition to, the ships' normal replenishment needs. As Quinn noted, "It's a very efficient way of doing business."[22]

Meanwhile, the Fleet and Industrial Supply Center provided manpower to assist with the logistics operation. According to Rear Admiral William Kowba, commanding the Fleet and Industrial Supply Centers, "Our FISCs were uniquely positioned and equipped for this humanitarian mission. We have a global presence and a trained team of military and civilian logistics professionals. The dedication and hard work displayed by our team at FISC San Diego, FISC Pearl Harbor and its detachment in Guam, and our forward FISC in Yokosuka, validated the COMFISCS vision."[23]

Through mid-January 2005, more than two thousand pallets of relief supplies were loaded in Singapore onto USNS *Concord* (T-AFS 5), *San Jose,* and *Niagara Falls* (T-AFS 3), as well as the fleet replenishment oiler *Tippecanoe,* for delivery to the frontline ships. Captain David Fitzgerald, the Naval Regional Contracting Center's commanding officer, said at the time, "It's been an all-hands effort from our military, civilian and host nation staff," purchasing relief supplies, storing them, inventorying, and then moving supplies to the ships and aircraft that would deliver them to the areas in need.[24]

Singapore was also designated the regional mail center for the operation. In addition to U.S. government supplies, individuals could also help, through the regular mail service. For example, a volunteer on the USNS *Mercy* asked a local doctor what he needed and immediately e-mailed the list to his wife back in the United States, who bought the items and put them in the mail.[25] The volume of mail skyrocketed from a norm of about ten thousand pounds of mail per fleet delivery to forty thousand. Singapore's post office staff of four quickly grew to fifteen. Early in the operation the carrier USS *Kitty Hawk* provided four carrier onboard-delivery aircraft to deliver supplies, mail, and personnel directly to *Abraham Lincoln.*[26]

Engineering and Rebuilding

While other U.S. services were delivering supplies, Seabees and the Army Corps of Engineers began the task of rebuilding basic infrastructure—roads, bridges, docks, and water, sewerage, and electrical services—which would help the local economy and markets to recover. Of the $950 million in aid that would eventually be pledged by the

United States, more than a third—$339 million—was "devoted to constructing roads, schools, water-distribution systems and other projects."[27]

Early in the mission, engineers on *Abraham Lincoln* responded to unusual logistical demands by designing a water manifold that filled containers automatically. Overnight, ship's technicians built a prototype that could fill seven to eight hundred five-gallon jugs per hour. Several other units were quickly built and given to other ships.[28] As Captain Raymond Ginetti put it, in Iraq the "bombs stacked up. . . . [H]ere the water stacked up."[29]

Relief efforts were hampered by the widespread destruction of roads, ports, and airfields, particularly in northern Sumatra: "There is nothing left, to speak of," said Lieutenant Commander Jeff Vorce: "Villages, one after the next, were obliterated. Concrete foundations were all that remained of most structures. Only a few mosques remained intact, surrounded by wasteland. Thousands of emerald green rice paddies had been peeled away, replaced by fetid swamps, mangled tree trunks and sea slime."[30] According to Senior Chief Jesse Cash: "In my 17 years of service, I have never seen such devastation, and I hope that I'll never see such again in my life." In Lam Jamek, road conditions were so poor that refugees used an elephant to pull a car to Banda Aceh.[31] For such reasons the roads and airports in northern Sumatra could not keep up with the logistical demands. To the south of Banda Aceh, in Medan, a virtual mountain of rice, instant noodles, and crackers sat waiting on the airfield, their delivery hampered by the small size of the airport and limited infrastructure, as well as by poor coordination and communications. Major Dwight Neeley, USMC, could only scratch his head when he saw a C-130 land at the adjacent commercial airport, far from the cargo and supply dumps at the military base: "This is the kind of stuff we have to deal with."[32]

Mosque—only remaining structure at Lhonga, Aceh Province.

By late January Martin Unternahrer, spokesman of the International Committee of the Red Cross, reported that Red Cross ships were on their way to Indonesia, where they would begin to bring supplies to tsunami victims. According to Col. Mark Schissler, commander of the Air Force forces forward and the expeditionary wing commander, the transition from the military to the Red Cross would be delicate: "The further from the crisis you get, you want to be more deliberate about how you distribute. They need to be fed systematically. [And] you have to support local markets."[33] Trevor Rowe, spokesman for the UN World Food Program, emphasized the point: "The important thing is that the United States military was right there at the beginning and made a huge difference. They had the logistical prowess . . . and without that we would not have been able to distribute to the remote areas."[34]

The engineering and repair work by ship's crew members helped to reopen Aceh Province's transportation system and commercial networks. As the mission passed the critical stage, local governments could once again undertake the task of rebuilding their own infrastructure, with UN support. Supplies were once again flowing normally into the area. To help people rebuild their lives, it was crucial for the region's long-range recovery that distribution of the goods be assumed by local markets.

Accidents and Logistical Bottlenecks

Senior Pentagon and military officials were especially concerned "to open up logistical bottlenecks and begin ferrying water, food, medical supplies and shelter."[35] Accordingly, a great many ships, people, and fixed-wing aircraft and helicopters moved supplies. Accidents were inevitable, and when something went wrong it could all too easily shut down the logistical flow. At least one fixed-wing airplane and one helicopter crashed during UNIFIED ASSISTANCE, and on both occasions only rapid response on the part of sailors and Marines kept the supply lines open.

One example of a potential logistics nightmare was the crash of a Tri-MG Boeing 737 into a water buffalo while landing at the Banda Aceh airport. Salvage engineers immediately contacted Boeing representatives, but the derelict plane had shut the airport down, and no further humanitarian supplies could be shipped in until it was moved.[36] The accident happened at night; early the next morning crewmen from *Abraham Lincoln* were flown in by helicopter to try to help move the Boeing off the airstrip. Their biggest problem was that none of the airport equipment could lift something so heavy. A UN-provided forklift was rated at thirty tons, but the aircraft weighed almost 150. Using a truck to drag the Boeing, two forklifts to steady it, and an airbag set that had been flown in from Singapore, the team from *Abraham Lincoln* finally managed to "crib" the plane off the runway. In the end, this task proved to be truly international; all the represented nations had to work together. As soon as the salvage teams had cleared the tarmac civilian aircraft started warming up, and twenty minutes later they began to take off.

The entire operation, once all necessary equipment had been assembled, lasted from three o'clock in the afternoon to 5:30, and by 6:40 the crews had managed to return to their ships in time for dinner. Since the Banda Aceh airport was the sole aerial port of debarkation, if the team had not been able to move the airplane by evening, no supplies could have come in that night. An American military spokesman, Captain Matt Klunder, told CNN, "Fortunately we had enough aid and supply gear that we were not set back, [but] tomorrow would have been an horrendous problem."[37] As three members of the *Abraham Lincoln* crew later recalled, "Using a motley collection of equipment—including an aircraft tractor, two forklifts, and a specialized dolly—they removed the airplane from the runway, reopening the sole lifeline of relief supplies into the region."[38]

Another logistical problem was the helicopter fleet itself, which grew to fifty-eight during the course of the operation. The aircraft were kept flying almost constantly, often loaded to capacity with personnel and supplies.[39] There was only one serious helicopter mishap, however: a SH-60 helicopter that crashed in a rice paddy adjacent to the airport at Banda Aceh.[40] When the crash occurred, on 11 January 2005 at 7:15 AM, hundreds of people, including crew members from the first C-17 to land at Banda Aceh airport that day, went running through the paddy to the downed chopper.[41] According to Captain Jeremy Boyd, thirty-three, a pilot of a KC-135R refueling craft and one of the first on the scene, a team of Australian medics even scaled a barbed-wire fence: "The Aussies did an outstanding job, really fast."[42]

After landing at Banda Aceh airport, cargo planes unloading at adjacent aid station.

A medevac team was immediately called in. No serious injuries had been sustained by the ten passengers and crew, but the helicopter was a total loss.[43] Helicopter pilots and crew took such occurrences in stride. According to Air Force master sergeant Richard Inman, forty, of Ithaca, New York, "The U.S. military is in the business of taking risks and that's one of them. It's an important operation bringing relief to hundreds of thousands of people."[44] Later, a board investigated the crash and determined that it had been due to a tail-rotor malfunction.[45]

Like the Boeing accident, the helicopter incident could have delayed supplies for days, since all airport operations had to be suspended.[46] The wreckage had to be recovered rapidly. Special salvage teams from *Abraham Lincoln* and *Bonhomme Richard* were sent to the site. First, the salvage team had to right the aircraft. *Bonhomme Richard* sent a Marine recovery team to do so; one Marine volunteer completely submerged himself in the mud to wrap a rope around the helicopter's tail. Meanwhile, members of the *Abraham Lincoln* crew made a helicopter sling, similar to those the Army used for recovering downed Blackhawks. Once the wrecked helicopter was upright and the sling put on, another helicopter picked it up and moved it to dry land, where it was reslung for the longer flight back to the ship.

Both of these accidents were relatively minor—remarkably so, considering the magnitude of the relief operation—but if not quickly resolved they might have produced logistical bottlenecks. That this did not happen was mainly due to the rapid response by U.S. military personnel. In both cases, once the wreckage was cleared away the airfield at Banda Aceh quickly resumed operations. Logistical lines were reestablished, and humanitarian supplies once again began to flow.

* * * * * * *

Throughout UNIFIED ASSISTANCE, the movement of supplies was the highest priority, with U.S. Navy ships and MSC prepositioned ships providing the first wave of water, food, and medicine. The air links from Utapao and air and sea links from Singapore came next; fixed-wing cargo aircraft, including C-130s, C-17s, and C-5s, flew 1,300-plus missions into the affected regions.[47] During this period, the movement of supplies was largely organized by *Abraham Lincoln*.[48] According to Lieutenant Commander Michael Hsu, "planning was for all militaries," but in this case "*Lincoln* had control."[49]

It was especially important to coordinate relief efforts with groups ashore. Every evening at seven o'clock a person from *Abraham Lincoln* attended a two-hour meeting to determine what supplies were needed and where they had to go on the following day.[50] Every morning on board the carrier the day's flight schedule was distributed outlining the whole day. Once everyone was in place, they started "to fly as many missions as was possible."[51]

Planned goals were often exceeded, after which operations became more like "freelancing."[52] As a result, Rear Admiral Crowder directed that execution be totally decentralized and empowered people to make decisions on the spot. In essence, he gave his crew a job and charged them to do it as "safely as possible."[53] Crowder's hands-off methods worked especially well when combined with the flexibility provided by helicopter-assisted access.

Notes

1. Epigraph quote in Deputy Secretary of Defense Paul D. Wolfowitz and Republic of Indonesia Minister of Defense Juwono Sudarsono, "Press Conference Transcript," Jakarta, Indonesia, 16 January 2005, available at www.pacom.mil/speeches/sst2005/050116 wolfowitz1.shtml.
2. Chris Seiple, *The U.S. Military/NGO Relationship in Humanitarian Interventions* (Carlisle, Pa.: U.S. Army War College, Peacekeeping Institute Center for Strategic Leadership, 1996).
3. "Operation UNIFIED ASSISTANCE: Naval Aviation's Swift Response to the Naval Disaster," *Naval Aviation News* (March–April 2005).
4. Commander Ted Williams, executive officer, VAQ-131, Naval Historical Center [hereafter NHC] Oral History, 20 February 2005.
5. Michelle Malkin, "A Salute to the USS *Abraham Lincoln*," Creaters.com, 5 January 2006.
6. Supply Department, CCSG-9, NHC Oral History, 20 February 2005.
7. Richard Spencer, "U.S. Marines Stage Beach Landings to Bring in Supplies," *London Daily Telegraph*, 11 January 2005.
8. CWO2 Leo Corpuz, USNS *Mercy* (T-AH 19), NHC Oral History, 4 June 2005.
9. Commander Jamie Garcia, USNS *Mercy* (T-AH 19), NHC Oral History, 4 June 2005.
10. Lieutenant Commander John Parker, USNS *Mercy* (T-AH 19), NHC Oral History, 5 June 2005.
11. Corpuz oral history.
12. Garcia oral history.
13. Robert Goodrich, "Relief Effort Dwarfs All for Scott Command, General Says," *St. Louis Post Dispatch*, 7 January 2005.
14. "Operation UNIFIED ASSISTANCE (OUA): Southeast Asia Earthquake and Tsunami Relief Effort (2004–2005)," Military Sealift Command report.
15. Lieutenant Commander T. S. Snider, "MSC Participation in Operation Unified Assistance (OUA)," 3 March 2005.
16. Fifi Kieschnick, "Ingleside Ship Deploys to Support Disaster Relief Efforts," *Navy NewsStand*, 5 January 2005.
17. Lieutenant Carrie Kimble, SC, USN, "*San Jose* Is a Vital Part of Operation UNIFIED ASSISTANCE," *Navy Supply Corps Newsletter*, May 2005.
18. Ibid.
19. Captain Jon Kaufmann, Commander, DesRon 9, NHC Oral History, 20 February 2005.
20. Lieutenant Commander Scott Murdock, DesRon 9 chief staff officer, NHC Oral History, 20 February 2005.
21. Lieutenant Chuck Bell, "Singapore Logistics Command Keeps Humanitarian Relief Flowing," *Navy NewsStand*, 17 January 2005, available at www.navy.mil/.
22. Ibid.
23. "Navy Logistics Supports Operation UNIFIED ASSISTANCE," *Navy Supply Corps Newsletter* (March/April 2005).
24. Ibid.
25. HM2 Jason Moore, USNS *Mercy* (T-AH 19), NHC Oral History, 5 June 2005. This example occurred during *Mercy*'s later humanitarian mission to New Guinea.
26. Commander Ron Hughes, strike operations officer, USS *Abraham Lincoln* (CVN 72), NHC Oral History, 20 February 2005.

27. Glenn Kessler, "White House to Seek $950 Million for Tsunami Aid," *Washington Post*, 10 February 2005.

28. Commander Chris Meyer, chief engineer, USS *Abraham Lincoln* (CVN 72), NHC Oral History, 20 February 2005.

29. Captain Raymond Ginetti, navigator, USS *Abraham Lincoln*, NHC Oral History, 20 February 2005.

30. Denis D. Gray, "U.S. Military Speeds Relief: Air Drops Offer View of Tsunami Destruction," *Washington Times*, 3 January 2005.

31. Denis D. Gray, "U.S. Choppers Find Remote Survivors: Additional Marines Head to Indo," *Washington Times*, 4 January 2005.

32. Richard C. Paddock, "Navy Airlifts Supplies to Ravaged Coast," *Los Angeles Times*, 3 January 2005.

33. Juliana Gittler, "U.S. Military Begins to Leave Tsunami-Stricken Region," *Pacific Stars and Stripes*, 24 January 2005.

34. Melanie Eversley, "Relief Groups Pick Up Where U.S. Leaves Off," *USA Today*, 25 January 2005.

35. Thom Shanker and James Brooke, "Tsunami Tests U.S. Forces' Logistics, but Gives Pentagon a Chance to Show a Human Face," *New York Times*, 9 January 2005.

36. Captain Matthew Klunder, DCAG, CVW-2, NHC Oral History, 20 February 2005.

37. "Water Buffalo Closes Aceh Airfield," AFP via Australian Broadcasting Corporation, 4 January 2005, available at www.command-post.org/nk/2_archives/018408.html.

38. Mark S. Leavitt, Jeffrey M. Vorce, and Michael M. Hsu, "For Compassion and Country: UNIFIED ASSISTANCE," U.S. Naval Institute *Proceedings* 131, no. 4 (April 2005), pp. 44–49.

39. Klunder oral history.

40. Jocelyn Gecker, "Copter Crash Adds to Tsunami Relief Challenges," *Philadelphia Inquirer*, 10 January 2005.

41. Rafael D. Frankel, "U.S. Cargo Plane Brings Lifesaving Supplies to Region," *New York Daily News*, 11 January 2005.

42. Rafael D. Frankel, "U.S. Soldiers Relish Their New Role: Lifesavers," *Boston Globe*, 12 January 2005.

43. Williams oral history.

44. Frankel, "U.S. Cargo Plane Brings Lifesaving Supplies to Region."

45. Commander Michael Horan, commanding officer, HS-2, NHC Oral History, 20 February 2005.

46. HSL-47, Group 1, NHC Oral History, 20 February 2005.

47. Ralph A. Cossa, president of Pacific Forum CSIS, "U.S. Military Provides 'Logistical Backbone' for Tsunami Relief," *eJournal USA: Foreign Policy Agenda* 9, no. 3 (November 2004), usinfo.state.gov/gi/Archive/2005/Mar/05-433015.html.

48. Lieutenant Commander Joel Moss, HS-2, Group 1, NHC Oral History, 20 February 2005.

49. Lieutenant Commander Michael Hsu, LNO, USS *Abraham Lincoln* (CVN 72), NHC Oral History, 20 February 2005.

50. Rear Admiral Douglas Crowder, CCSG-9, NHC Oral History, 20 February 2005.

51. Hsu oral history.

52. Ibid.

53. Crowder oral history.

CHAPTER FIVE

U.S. Navy On-Site Delivery
Helicopter-Assisted Access

Helicopters are invaluable, especially helicopters coming in from the sea, where they can be refueled and resupplied out on our carriers, and are not taking up space at airfields or putting a logistics base at airfields.

COLIN POWELL, U.S. SECRETARY OF STATE, 3 JANUARY 2005

Sea basing proved essential both to organize and later to coordinate the arrival of humanitarian supplies.[1] However, as Rear Admiral Crowder acknowledged, no matter where the supplies came from, "Helicopters provided [the] last link in the chain as the only way to get the supplies to those in need." When the helicopters went inland, they would first locate survivors. Landing on an undamaged section of road or in what was left of a village square, they would hand out supplies as quickly as possible before being completely swamped by people. Five or six days after the tsunami, some of the victims had become so desperate that they would rush even airborne helicopters.[2]

As later events would show, the sequencing of the arrival of warships proved extremely important, with *Abraham Lincoln* arriving first, followed several days later by *Bonhomme Richard,* and later still by *Essex.* The crew of *Abraham Lincoln* was most concerned with providing immediate aid, especially food and water, in the hardest-hit areas of Aceh Province, leaving *Bonhomme Richard* and *Essex* to focus on the most efficient dispersal of supplies.[3] The exact methods utilized to make deliveries were generally left to those on the spot.[4] But as Rear Admiral Crowder explained, because so many bridges had been destroyed and many major roads were impassible, the "missing element" was what *Abraham Lincoln* could provide—helicopters.[5]

In addition to the cooperation of all the ships under the commanders embarked respectively on board *Abraham Lincoln, Bonhomme Richard,* and *Essex,* the mission in northern Sumatra also required coordination with civil affairs groups (CAGs) in Indonesia, a flow of materials through Thailand and Singapore, and on-site, day-to-day, sea-based command provided by Rear Admiral Crowder on *Abraham Lincoln.* Finally, and most importantly, it required direct air access from the ships to the shore, which was made possible only by the fact that *Abraham Lincoln* was carrying extra Seahawk helicopters.

The B2C Helicopter Program

Abraham Lincoln had been involved in a proof-of-concept program, known as "Bravo to Sea," or B2C, designed to improve the integration of helicopters into carrier flight operations. As a result, the carrier had an unusually large number of helicopters on board. Its helicopter crews were largely integrated into the air wing and had received additional flight training in carrier operations.[6] The program was part of Navywide preparations for a rotary-wing operational reorganization.[7] In September 2004, three months before the tsunami disaster, technicians had tested *Abraham Lincoln* for static electricity discharges that could adversely affect helicopters during refueling. In early December 2004, further tests helped the refueling needs of the additional helicopters. *Abraham Lincoln* was the first carrier strike group to deploy with a light helicopter antisubmarine squadron, HSL-47, with eight capable H-60 Seahawks in its air wing. This was more HSL aircraft than in "any other strike group prior to this."[8] By using helicopters intead of S-3 tankers as antisubmarine warfare platforms, the ship "went from 7 to 17 helicopters. OUA made it a perfect match."[9]

After the tsunamis hit, the Indonesian government had only two helicopters on the entire island of Sumatra.[10] Later, the USS *Bonhomme Richard* brought another twenty-five helicopters, including CH-53s and CH-46s. Air Force C-17s also quickly delivered six HH-60 helicopters and two more CH-46 helicopters.

As UNIFIED ASSISTANCE continued, it became clear that "*Lincoln* Strike Group was perfect for this mission as extra helicopters and pilots were embarked."[11] Indonesian military spokesman Ahmad Yani Basuki complimented the helicopter access: "They've helped us reach places we have not had the time, or manpower, or equipment to go to. It really speeds up the distribution of aid."[12]

Aceh Province

The geography of northern Sumatra made sea basing, forward sea-based logistics, and especially helicopter access crucial. The earthquake and tsunami damage was most extreme in this region, since it had been closest to the seismic epicenter. The loss of communications between the rural areas and the capital at Banda Aceh was almost total, and for several days there were no reports. All of the land routes were damaged, and many were completely impassable; the survivors could be assisted only by air. Further, air access from the sea proved particularly useful, since along Aceh Province's northwestern shoreline is a largely impassable mountain range. Only a single north-south road served the entire coastline from the northern tip to the coastal city of

Improvised landing zone at Kreung Raya, Aceh Province.

Meulaboh, where a small road cut east across mountains to the larger city of Medan. When the earthquake and tsunamis almost completely destroyed the coastal road, over a hundred miles of coastline were cut off from the interior.[13]

Even before *Abraham Lincoln* arrived, Rear Admiral Crowder had ordered his Civil Affairs Group to coordinate with the Indonesian authorities. Once the ship reached Banda Aceh, the group was the first on the ground, making contact with local government officials. The governor of Aceh Province was supposed to be in touch with local representatives in Banda Aceh, but in practice communications were poor, and "helicopters arrived with no one to unload the supplies." Eventually, Indonesian general Bambang Darmono ended up personally coordinating the effort, "so things started getting done in a way that was constructive."[14] Meanwhile, Ambassador B. Lynn Pascoe and the American embassy in Jakarta became an important "hub of all information," and were responsible for direct coordination with the Indonesian government.[15]

The first helicopter-related mission facing *Abraham Lincoln* entailed scouting out the area and linking up with the local and other foreign aid workers. Commander Ted Williams, a fixed-wing pilot, became liaison officer to the foreign aid providers. Beginning on 1 January 2005, for example, he worked with Australians at Banda Aceh airport to overcome difficulties delivering supplies from an Australian C-130 to outlying areas.[16] Over time the number of foreign nations represented increased, making liaison work more and more important.

The first day of the mission was perhaps the busiest: locating appropriate landing zones, determining how many boxes could be carried on each flight, and setting up the landing zones. In order to ensure a rapid learning curve, the same officer and crew remained together throughout this ramp-up phase.[17] As Commander Dan Boyles recalled, "With all the devastation, the mosques appeared to be the only thing left standing."[18] Therefore, mosques often became the focus for helicopters trying to find landing sites to distribute water and food, although sports fields, sections of roadway, dry fields, and even concrete home foundations were used as landing sites.[19]

It was important that the mission be both joint and combined, and that Indonesian representatives be acknowledged clearly as in charge. When flying into a new area, each helicopter would take a member of the Tentara Nasional Indonesia, the Indonesian military, and leave him to organize survivors to unload supplies.[20] There was a conscious effort to make even loading supplies onto helicopters ashore a combined effort. Boxes would be passed down a long line like fire buckets, with the men positioned by nationality, "one Indonesian soldier, then one American."[21]

Efficient distribution also required understanding the local geography. Each evening the Indonesian officials in Banda Aceh created and distributed an air tasking order listing supplies and passengers to be moved, the numbers and types of helicopters needed for particular tasks, and the landing locations. Often the landing points were described only by latitude and longitude; sometimes pilots would report, "That location is underwater," so enormous had been the topographical changes caused by the disaster.[22]

Developing a New Flight Routine

Every morning at sunrise the Seahawks took off from the carrier with shore teams for Banda Aceh, returning only to refuel or swap crews, which they were required to do every eight hours. Most of the day the helicopters were kept busy flying supplies to outlying areas, picking up stragglers on the return trips and flying them directly to Banda Aceh. According to Lieutenant Scott Cohick, one of the Seahawk pilots, the damage was overwhelming: "You see these places that used to be villages. And now there's only a mosque and lines that used to be streets."[23]

The first supply helicopter mission from *Abraham Lincoln* took place on 2 January 2005. In response to the "desperation on faces of survivors," one helicopter squadron increased its flight tempo, flying "4 missions first day, then flew all aircraft until they couldn't be flown any more." Hangar crews had to work especially hard to keep the helicopters functioning properly; Commander Michael Horan, commanding Helicopter Antisubmarine

Squadron 2, commented in particular on the "herculean effort from maintenance crews."[24] According to Commander Jon Ross, this meant that everyone worked three-day rotations, amounting in practice to little more than "organized chaos."[25]

Helicopter pilots later recounted both the joys and the difficulties of this new flight schedule. Commander Boyles flew about seventy hours total, while the air wing accounted for 1,200 hours of flying time, equivalent to the flight time in the first three months of a normal deployment. For many, however, it was a dream come true: Boyles considered the relief mission the "most fun flying in 18 years, no rules . . . no airspace issues, only our own aircraft deconfliction. License to fly within safety limits of plane."[26]

AM3 Chad Gillies (HC-11 Det 2).

It was not all fun and games, however. Crowds often rushed the helicopters as they landed, and if the helicopter crews were not careful, men would overpower women and children to get to the food and water.[27] During aid drops in Lamno and Calang, villagers swarmed the helicopter, forcing the pilot and crew to dump the boxes of aid as fast they could: "We have to close the windows or they'll climb through them. They'll come right in. They'll climb on the helicopter."[28] According to Commander Boyles, the dangers of landing near desperate refugees caused the missions to be changed; helicopters began to fly to areas with fewer people "but more crowd control."[29] Communications with refugees was initially difficult, because there were not enough translators to explain the dangers of helicopters.[30] Later, local translators were assigned by the Indonesian government to fly with each helicopter. They used large cardboard placards to warn refugees away from the rotor blade dangers.[31] The translators would also tell the crowds to back away so that the helicopter could land safely.[32]

The delivery of fresh water remained one of the highest priorities throughout the operation. Food can be packaged and dropped from the air, but water, because of its weight, cannot be without bursting its containers. But food without water merely leads to dehydration, so the helicopters had to transport both. Helicopter crews would set aside boxes of bottled water in case they ran across isolated refugees. If they did, they would land at the next suitable location, fly back to the refugees, and direct them to where the water had been placed.[33]

To begin with, the helicopter crews carried whatever food rations the ship had to spare. As the logistical train from Utapao and Singapore became more reliable, requests for types of food became more specific. For example, the Indonesians politely declined *Abraham Lincoln*'s offer of canned fish, probably because it was packed in unfamiliar oils or tomato paste and so was considered difficult to incorporate into the locals' regular diet. In any case, within a couple of days of the tsunami fishermen were once again going out to sea and fresh fish became more widely available.[34]

As soon as the mission got well under way, the airport at Banda Aceh became a primary hub of the relief operations ashore. Fixed-wing aircraft, including C-130s and C-17s, would bring in supplies, which would then be offloaded into waiting trucks. Trucks transported the supplies to the helicopter landing zones, where "lines of sailors transferred supplies from trucks into waiting helicopters as the pilots determined the maximum cargo they could carry to their assigned destinations."[35]

People desperate for water, Keudeteunom, Aceh Province.

Helicopters were not initially allowed to pick up refugees; the Indonesian government wanted to keep additional injured out of the overcrowded capital. But early in the operation pilots began to disregard the government demand not to rescue injured villagers.[36] Locating and airlifting injured Indonesians to the hospital at Banda Aceh became an important task of helicopter pilots and their crews. To prevent overloading the civilian facilities, however, a field station staffed by Navy medical personnel was hastily set up at the airport. Triage separated those who needed to be transferred to hospitals.[37] One injured boy brought in by helicopter was vomiting sand, proof that he had been submerged for an extended period. Lieutenant Lisa Peterson, a doctor from *Abraham Lincoln,* reported, "We are seeing a lot of broken legs, a lot of lacerations, a lot of pneumonia from all the salt water—about three quarters of the patients have pneumonia, among other things."[38]

Sea-based helicopters proved absolutely crucial in delivering humanitarian supplies and moving injured people to hospitals. Helicopters were the only way to move supplies efficiently from the logistical ports of entry at Banda Aceh and Meulaboh to the "more than 60 villages and camps of displaced persons along the coast." In the early stages of the operation, these tasks fell primarily on *Abraham Lincoln*'s seventeen helicopters: "While only a few of these aircraft were specifically designed for logistics, they served capably in their life-saving role."[39]

Vertical replenishment from the USNS *Rainier* to Banda Aceh airport.

The Impact of Air Access on the Media and NGOs

Sea basing and helicopter access helped the media and international aid groups carry out their work in northern Indonesia. *Abraham Lincoln* prepared in advance to receive the press, having an additional public affairs officer (PAO) assigned temporarily from the carrier USS *Kitty Hawk* (CV 63) to assist.[40] This PAO explained the process of working with the media and nongovernmental organizations (NGOs) and helped delegate their representatives to particular helicopters. The arrival of the media in particular could not help but affect the primary humanitarian mission; shuttling the news teams back and forth did not compromise the mission, but it did significantly reduce the number of helicopter flights available to deliver supplies.[41] The trade-off, from the Indonesians' point of view at least, was that international donations and humanitarian aid grew quickly as televised pictures of the tsunami damage appeared around the world.

The appearance of NGOs and other international groups sometimes threatened to undermine the smooth working of the relief mission. Reportedly, some 109 such groups were eventually represented in Banda Aceh, and there was significant "rivalry among NGOs to try to get aid to everyone."[42] According to one pilot, "Lots of media and politics involved so that became more aggravating to helo crews. Media in general were not bad but they took up space."[43]

An additional PAO was soon flown in to help with the media and distinguished visitors. Normally PAOs handled all press interactions, but there were so many that this proved difficult.[44] Lieutenant Commander Jason Salata, a Public Affairs Officer attached to the American Embassy in Jakarta, emphasized the importance of making a U.S. Navy PAO available to work with the embassy early in the mission, to help answer military questions and interact directly with other PAOs assigned to the disaster area.[45]

As a matter of policy, the U.S. Navy went out of its way to assist the media in any way that it could. CNN may have broadcast the images and reports, but few reporters could have arrived on location to collect them without the help of the helicopters.[46] According to Lieutenant Commander John Daniels, the media rush began on 2 January 2005, when Dan Rather showed up with his *60 Minutes* team. To begin with, the international media had little choice but to use footage prepared in advance by the Navy PAOs: "Some of the first video was shown by CNN and CBS from that shot by the crew."[47]

Aboard *Abraham Lincoln,* many of the crew members who could not go ashore did their part by assisting the press, health assessment teams, or United Nations coordinators by providing berths, laundry service, and food. In particular, *Abraham Lincoln* assisted the UN assessment teams traveling around Sumatra to determine the extent of the damage. According to Rear Admiral Crowder, the UN assessment as to which areas

Rear Admiral Douglas Crowder, *Lincoln* strike group commander, with Dan Rather, CBS News anchor.

in Sumatra required long-term assistance was one of the most important accomplishments of the entire operation.[48]

Sailors were "embedded" with civilian representatives of NGOs and the UN. They would provide escort to the helicopters, arrange rooms for charts, and assist with setting up "photo ops." To many crew members of *Abraham Lincoln* it seemed like a huge publicity coup for the Navy, since these outside groups provided a "higher level of awareness" about how well OUA was progressing.[49] Very soon, images of American helicopters delivering supplies became a regular feature on Indonesian television. This positive press helped change local attitudes toward the United States "for the better."[50] One poll conducted by the Indonesian Survey Institute in February 2005 concluded that 65 percent of Indonesians now had a more favorable view of the United States.[51]

Nonetheless, and although working with them was mostly a positive experience, the sheer numbers of NGOs and other international groups trying to help Aceh Province threatened to undermine the relief mission. Australia decided not to work with them at all but only with the host country in disbursing its aid. According to the Australian foreign minister, Alexander Downer, "That is a much better outcome than pushing money through international organizations."[52] The U.S. government, however, welcomed the participation of NGOs and other aid groups, generally maintaining good relations with them, on the theory that when the U.S. Navy withdrew private groups could help pick up the slack. To assist the NGOs in their interaction both with the U.S. military and with each other, the crew of *Abraham Lincoln* created a management information center in a rented house in Banda Aceh.[53] This center assisted the aid organizations to find out what was happening locally, regionally, and in other areas affected by the tsunamis. It also helped relieve the helicopter pilots and crews of constant demands to assist the NGOs.

High Stress Levels

The helicopter pilots and their crews were faced with an enormous task. Commander Frank Michael later stated that the January flying average was twenty hours more per pilot than normal, effectively similar to wartime: "Take in supplies, load up victims, do debrief later."

Fire brigade–style loading of supplies by Indonesians and Americans, Aceh Province.

To keep up the tempo, they "agreed to fly all helos all the time and only layup as needed." The result: "flight deck crews worked 24/7, usually 12–14 hour days each."[54]

During the first two weeks of the mission, the weather was especially bad. Under normal conditions, helicopters would probably not have been allowed to fly at all, but the missions came first.[55] Day after day, week after week, the helicopters were in the air on a daily basis moving personnel to and from shore, carrying much-needed supplies to those areas that needed them the most, and taking sick and injured to hospitals.

Helicopter crews were usually unarmed and often far from their sea bases. One pilot delivering supplies thought that a disgruntled Indonesian was attacking his helicopter with an upraised machete but then realized that the man was merely hacking through brush and debris to reach the aircraft.[56] Once an Indonesian soldier firing into the air in an attempt to control two dozen refugees scrambling for supplies almost hit a helicopter's rotor blades.[57] Even children could be a problem; some devised a dangerous game of jumping up close to the moving rotors and allowing themselves to be blown backward on the rotor wash.[58]

The normal routine became for each helicopter first to fly several trips carrying passengers from the ship to Banda Aceh. Once all of the shore working party was in place, the helicopters would conduct humanitarian flights for the rest of the day. Concerns with security and the lack of proper housing facilities meant that the helicopters had to bring everyone back to the ship; nobody stayed ashore at night.[59] Helicopter-capable ships handled as many as ninety landings per day, mainly to refuel helicopters, and aircraft maintenance crews worked long hours in hot and humid conditions.

For the helicopter pilots, stress levels were high, especially in developing appropriate landing and take-off procedures on land. According to Lieutenant Commander Joel Moss, "Challenges included following rules verbatim or not following rules" at all. Once a person on the ground threw an empty box into rotor blades just to see it being torn to shreds, which could have seriously damaged the helicopter. One helicopter was forced to lift off with desperate survivors hanging on; the pilot was "concerned over safety of crew so [the] choice was to lift off."[60] Faced with the possibility of damage and rightly concerned about the safety of the people below, sometimes pilots decided that it was not safe to land. Because they were not equipped with hoists to lower supplies to the ground, this meant that supplies were not delivered. It was several days before the desperation of the survivors lessened to the point where helicopter missions could become routine.

Because the physical limitations of sea basing made helicopter access absolutely essential, and in particular because of the enormous volume of logistical supplies that had to be transported, the helicopter pilots and crews experienced some of the highest workloads of any personnel involved in UNIFIED ASSISTANCE. Nonetheless, the majority of the helicopter pilots were later to recount how remarkable the experience had been for them. As Lieutenant Commander Jason Carter put it, OUA was a "typical navy response, as [we] weren't sure what the mission would be but we'd do it when we got there."[61]

* * * * * * *

Direct sea-to-land access provided by helicopters during operation UNIFIED ASSISTANCE made the fundamental difference.[62] Because of the B2C program, *Abraham Lincoln* arrived off Aceh Province with a much stronger helicopter force than normal; several days later, this was augmented by the even more robust helicopter fleet of

Helicopter crew members looking for suitable landing zone, Aceh Province.

Bonhomme Richard. Former Indonesian prime minister and social welfare minister Alwi Shihab, observing a line of U.S. Navy Seahawks carrying food and water to victims of the tsunami, declared, "Isn't it incredible?"[63] Even so, the first days of the relief mission were quite difficult, especially for the helicopters and their pilots. The U.S. Navy had to coordinate with the local government, PACOM, task force headquarters in Utapao, and many other foreign countries as well. With many international bodies competing to assist the tsunami victims, it was to prove particularly difficult to agree on any one set of rules or operating procedures.

According to Commander Ron Hughes: "No JFAC [Joint Force Air Component] [was] needed but this was a CFAC [Combined Force Air Component]."[64] Coordination with Australian, Singaporean, Japanese, German, and French ships grew quickly; foreign navies were added to the coalition and furthered the humanitarian mission in numerous ways. *Abraham Lincoln* assisted by refueling helicopters; also, for periods of twenty-four hours at a time, the ship was "designated 'ready deck' and could not do maintenance."[65] Coordination between the different ships was carried out mainly by phone and e-mail.[66] However, there were so many ships with helicopter assets that "cognizance of all helos was imperative," especially when the foreign helicopters were "allowed to land on the ships" to obtain fuel.[67]

Perhaps it was more a "collection" than a "coalition," but there was close cooperation nonetheless. During one episode a medical team included American pilots and medics, an Australian coordinator, doctors from China, and Indonesian soldiers, all working side by side. The international language was often the "thumbs up," as two "Indonesians, an American and an Australian held the four corners of a stretcher with an injured boy and lifted it carefully from one of the helicopters."[68] Reliable ship-based communications proved to be the best way to integrate all of the many countries and their disparate missions.

Notes

1. Epigraph quoted in Donna Leinwand, "Helicopters Help Relief Operations Gain Momentum: Can Fly Where Bridges, Roads Are Washed Out," *USA Today*, 3 January 2005.
2. Rear Admiral Douglas Crowder, CCSG-9, Naval Historical Center [hereafter NHC] Oral History, 20 February 2005.
3. Captain Kevin Campbell, CCSG-9 staff N5, NHC Oral History, 20 February 2005.
4. Commander Russ Thompson, executive officer, HS-2, NHC Oral History, 20 February 2005.
5. Crowder oral history.
6. Commander Dan Boyles, executive officer, HSL 47, NHC Oral History, 20 February 2005.
7. Mark S. Leavitt, Jeffrey M. Vorce, and Michael M. Hsu, "For Compassion and Country:

UNIFIED ASSISTANCE," U.S. Naval Institute *Proceedings* 131, no. 4 (April 2005).

8. Crowder oral history.
9. Captain Matthew Klunder, DCAG, CVW-2, NHC Oral History, 20 February 2005.
10. Denis D. Gray, "U.S. Choppers Find Remote Survivors: Additional Marines Head to Indo," *Washington Times*, 4 January 2005.
11. Captain Jon Kaufmann, Commander, DesRon 9, NHC Oral History, 20 February 2005.
12. Gray, "U.S. Choppers Find Remote Survivors: Additional Marines Head to Indo."
13. Leavitt, Vorce, and Hsu, "For Compassion and Country."
14. Boyles oral history.
15. Commander Ted Williams, executive officer, VAQ-131, NHC Oral History, 20 February 2005.
16. Ibid.
17. Commander Jon Ross, executive officer, VFA-151, NHC Oral History, 20 February 2005.
18. Boyles oral history.
19. Leavitt, Vorce, and Hsu, "For Compassion and Country."
20. Wayne Arnold, "In Areas All but Unreachable, Helicopter Crews Bring Help," *New York Times*, 5 January 2005.
21. Eric Unmacht, "View from U.S. Chopper," *Christian Science Monitor*, 6 January 2005.
22. Leavitt, Vorce, and Hsu, "For Compassion and Country."
23. Arnold, "In Areas All but Unreachable, Helicopter Crews Bring Help."
24. Commander Michael Horan, commanding officer, HS-2, NHC Oral History, 20 February 2005.
25. Ross oral history. The USS *Rainier* had HC-11, a helicopter logistics squadron, which included two MH-60s specially configured to move supplies. Together they logged 363 hours of flight time during one month, and Lieutenant Commander Chris Hayes flew 140 hours.
26. Boyles oral history.
27. HS-2, Group 1, NHC Oral History, 20 February 2005.
28. Alan Sipress and Glenn Kessler, "A Clear Line between Life and Death," Washington Post Foreign Service, 5 January 2005.
29. Boyles oral history.
30. HSL 47, Group 2, NHC Oral History, 20 February 2005.
31. Lieutenant Commander Tony Ohl, CCSG-9 staff N5A, NHC Oral History, 20 February 2005.
32. HSL 47, Group 2, oral history.
33. Sipress and Kessler, "A Clear Line between Life and Death."
34. "Report of Listening Project, Aceh, Indonesia" November 2005, available at www.cdainc.com.
35. Horan oral history.
36. Donald G. McNeil Jr., "Hearts, Minds and Flood Relief," *New York Times*, 9 January 2005.
37. Paul Wiseman, "U.S. Troops Report Successes in Relief Effort: Governments around World Sending Help," *USA Today*, 12 January 2005.
38. Jane Perlez, "Pace of Global Aid Effort Picks Up on Indonesian Coast," *New York Times*, 4 January 2005.
39. Leavitt, Vorce, and Hsu, "For Compassion and Country."
40. Ross oral history.
41. Ibid.
42. Commander Dan Verheul, N2/ACOS, CCSG-9, NHC Oral History, 20 February 2005.
43. HSL 47, Group 2, oral history.
44. Captain Bob Aronson, CCSG-9 staff N6/ACOS C41, NHC Oral History, 15 February 2005.
45. Lieutenant Commander Jason Salata, PAO, PACOM, Pearl Harbor, Hawaii, NHC Oral History, 2 March 2005.
46. Williams oral history.
47. Lieutenant Commander John Daniels, PAO, USS *Abraham Lincoln* (CVN 72), NHC Oral History, 20 February 2005.
48. Crowder oral history.
49. Commander John Baca, gunnery officer and weapons officer, USS *Abraham Lincoln* (CVN 72), NHC Oral History, 20 February 2005.

50. Neil King, Jr., and Donald Greenlees, "Aid Effort in Indonesia Could Lift U.S. Image in Eyes of Muslims," *Wall Street Journal*, 5 January 2005.
51. "Indonesia's Stature Rises with New Security Pact," *Christian Science Monitor*, 17 January 2006.
52. David R. Sands, "U.N. Takes Charge of Effort to Help Tsunami Victims," *Washington Times*, 7 January 2005.
53. Verheul oral history.
54. Commander Frank Michael, HSL-47, commanding officer, NHC Oral History, 20 February 2005.
55. Horan oral history.
56. Wayne Arnold, "Satisfying Job for the Navy: Saving Lives of Survivors," *New York Times*, 6 January 2005.
57. "U.S. Military Ending Tsunami Relief Effort: Aid Groups Pledge to Bear Greater Share of Burden as Troops Begin Withdrawal," *Baltimore Sun*, 21 January 2005.
58. Arnold, "Satisfying Job for the Navy."
59. Lieutenant Commander Joel Moss, HS-2, Group 1, NHC Oral History, 20 February 2005.
60. Ibid.
61. Lieutenant Commander Jason Carter, CCSG-9 staff communications officer, NHC Oral History, 20 February 2005.
62. Ralph A. Cossa, president of Pacific Forum CSIS, "U.S. Military Provides 'Logistical Backbone' for Tsunami Relief," *eJournal USA: Foreign Policy Agenda* 9, no. 3 (November 2004), usinfo.state.gov/gi/Archive/2005/Mar/05-433015.html.
63. Alan Sipress and Noor Huda Ismail, "Relief Transcends U.S.-Indonesia Divide: Rights Concerns Sidestepped as Militaries Respond Jointly to Disaster," *Washington Post*, 4 January 2005.
64. Commander Ron Hughes, strike operations officer, USS *Abraham Lincoln* (CVN72), NHC Oral History, 20 February 2005.
65. Kaufmann oral history.
66. Carter oral history.
67. Hughes oral history.
68. Perlez, "Pace of Global Aid Effort Picks Up on Indonesian Coast."

CHAPTER SIX

Communications at Sea
Integrating the Collective

We've got a bizillion dollars of satellite communications suite here in the strike group, but we are not so good with how to communicate from the ship, where the helicopters are, to the beach, where the supplies are, three miles away from each other.

REAR ADMIRAL DOUGLAS CROWDER, *LINCOLN* STRIKE GROUP COMMANDER
20 FEBRUARY 2005

In addition to sea basing, logistics, and helicopter access, communications was vital, both vertically up the chain of command and horizontally among the commands and units in the theater.[1] Throughout UNIFIED ASSISTANCE, Lieutenant General Blackman warned that reliable communications was key to the success of the operation. He compared the "fog of relief" to the "fog of war," emphasizing "strong communication among all parties" as the only possible solution.[2]

Efficient communications systems, then, were essential, especially among the various countries cooperating to provide aid. On the ground, small groups known as "spark teams" provided communications between the ship and shore and among various shore parties. According to Commander Dan Verheul, the spark teams helped "provide the energy to make things happen."[3] Described by one source as a "clearinghouse for information," they also helped to overcome differences in hardware, software, and bandwidth between military and civilian participants.[4] This was particularly important since international aid workers, military personnel, volunteers, and government officials were all competing for access into the area.[5] The ability of these small teams to coordinate communications on the ground helped make the humanitarian effort a success.

For many sailors in USS *Abraham Lincoln* there seemed to be little or no reason to direct mission-oriented information first to Utapao and only then to Pacific Command; in any case, Utapao was much harder to contact than Honolulu. Keeping in touch with PACOM proved to be relatively easy, especially via the Navy's satellite e-mail network, but communications were shaky both within the chain of command and between the different nodes of the command and control structure.[6] It took many days, in some cases weeks, to fix all the regional communications problems.

Difficulties with Ship-to-Shore Communications

When *Abraham Lincoln* arrived off Aceh Province, ship-to-shore communications became especially critical to the success of the operation and so proved to be the "biggest challenge."[7] Once proper landing zones had been located ashore, crew members rigged radio communications with the ship. One of the greatest challenges for the crisis action teams was maintaining constant communications with the personnel working in Banda Aceh. Often it was easiest to use e-mail to communicate with the "away" teams, but for a number of reasons some locations could "not receive data."[8]

Commander Ted Williams later recalled that there were "no comms [on the] first day. Handheld radios came second day so things worked much better."[9] The Marines helped solve this problem by providing their cell phones.[10] Motorola walkie-talkies used by the explosive ordnance demolition teams proved especially useful, since they could function as ship-to-shore communicators. Munitions disposal in connection with humanitarian missions had been "non-existent before this operation. Then they became invaluable to the missions."[11] However, not only did their HF communications have to be "in-tune with weather" but rotor wash from helicopters could take down antennas and disrupt signals.[12]

Early in the mission, it was suggested that acquiring cell phones might provide the easiest solution. A team was sent in to perform a site survey but found that it would take too long to get a new system operating.[13] The Nortel Company at one point promised free cell phones, but the Navy's Judge Advocate General considered them "gifts" that could not be accepted.[14] Iridium satellite phones were also an option, albeit more expensive and difficult to use.

Abraham Lincoln spent most of the time on the west coast, from where the high coastal mountains blocked direct communications with the east side. To overcome this handicap, strike group ships were spread up and down along the coast to act as communications relays.[15] There was some horizontal communication and cross-talk among *Abraham Lincoln, Bonhomme Richard, Fort McHenry,* and *Essex,* when the latter arrived, but each had its own zone so that there was relatively little need for coordination. Due to "strike group–centric" thinking, there was not enough integration among different assets in the theater.[16]

The amphibious assault ship USS *Bonhomme Richard* off the coast of Aceh Province.

From time to time, *Abraham Lincoln* also developed serious satellite communication difficulties, especially when the satellite was directly overhead and an antenna experienced "block zones." Overall, however, satellite communications were reliable, and the shore parties "couldn't have managed without Satcoms and portable communications equipment."[17] The availability of global communications was to have an especially significant effect. As Kevin Parra noted, "Satcoms was the mission. Control element for ship and coordination of flights from ships, tracking food, medevacs, staging area. Backpacks were used and batteries were backups."[18]

To resolve communication and other problems, every night a meeting, called a "hot wash," was chaired by Rear Admiral Crowder or the deputy carrier air group commander, Captain Matthew Klunder. One issue addressed by these meetings, according to Captain Klunder, was difficulty in scheduling, including delays getting the work parties ashore or back to the ship.[19] Another was keeping track of personnel ashore; all volunteers were put into a database and assignments were distributed so "all aboard ship had an opportunity to participate."[20]

SIPRNET versus NIPRNET versus CENTRIXS Problems

One constant problem during UNIFIED ASSISTANCE was that the U.S. Navy ships were connected to the classified SIPRNET, not the unclassified NIPRNET (essentially the civilian Internet), which CSF 536 headquarters in Utapao was initially set up to use. One former submariner ruefully described how difficult it was to go from a ship that was trying to be stealthy to one that was trying to transmit as much as possible.[21] Lieutenant General Blackman wanted to keep everything unclassified, so he could work easily with coalition partners and with NGOs. This cut Utapao out of much of the early decision making by the naval forces.

The commanding officer and embarked staff in *Abraham Lincoln* faced a difficult problem. An aircraft carrier is "geared to work on the Secret level, and is not set up to work with unclassified" material. It is difficult to downgrade material from secret level to unclassified, so much crucial information appeared only at the secret level. Utapao was "cut out from all of this information" and so often operated "in the dark." As an expedient, most of the flag staff members were using unclassified e-mail accounts, which were not safe from an operational security perspective. Utapao acquired SIPRNET on 7 January, but it did not have many accounts. Also, its SIPRNET computers were housed separately from the rest, and there were not many of them, with the result that people did not monitor them "twenty-four/seven."[22]

To communicate from ship to ship, the top choice was chat rooms, which used SIPRNET almost exclusively. No similar architecture was available on NIPRNET, which made it more difficult to communicate with foreign navies. Also, the cost of using SIPRNET was high; it would normally be considered "too expensive" to use as a chat server.[23] Further, the U.S. Navy often used Internet Relay Chat, which was text based and often left Utapao out of the loop. Utapao needed to be able to use the same chat rooms as the ships but for several days nobody from Utapao was even visiting them.

One prime example of the disconnect between Utapao and *Abraham Lincoln* was the question of what to do with *Bonhomme Richard* when it arrived in the area. Rear Admiral Crowder's Joint Planning Group originally planned to use amphibious ships to set up a second sea-based logistical hub farther south, near Mulaboh. From there, it could use its helicopter force to transfer supplies. This would have been an efficient use of air assets. However, Utapao countermanded this plan and sent *Bonhomme Richard* instead to Medan, where it quickly overloaded the airport, virtually shutting it down. Only a few days later was *Bonhomme Richard* redirected to Mulaboh.[24]

Offload at Lamno, Aceh Province.

A second major concern was communications among different members of the coalition. According to Ralph A. Cossa, president of the Pacific Forum CSIS in Honolulu, country-to-country communications at the very top worked well: "The relief operations also demonstrated the merits of the Pacific Command's Asia Pacific Area Network (APAN). The APAN mission is to share unclassified information electronically in order to facilitate regional understanding, build confidence among Asia-Pacific neighbors and enhance security cooperation. APAN was used extensively by Britain, Canada, Australia and the affected Asian nations while coordinating relief efforts."[25] Within the region, however, the U.S. naval forces had to coordinate not only with each other but with other foreign aid countries. Additional helicopters were soon provided by Britain, Switzerland, France, Japan, Germany, and Malaysia. Australian air traffic controllers helped direct the more than 350 helicopter landings at Banda Aceh airport each day.[26]

Although a "combined coordination cell" was created to coordinate the twelve main countries involved, including Thailand, connectivity remained the "biggest challenge."[27] Had there had been a formal coalition, the Japanese and Australians would have wanted to use the Collaborative Enterprise Regional Information Exchange System (CENTRIXS) to communicate. But as it was, nobody wanted to use it, since it was difficult to operate and far too many participants did not have CENTRIXS terminals. The Global War on Terror "enclave" might have been better for this type of mission, but it could not be used for OUA. What was needed instead was a general HA/DR enclave for CENTRIXS, which would have made the operation look more like a coalition. In the end, it did not really matter, since "it wasn't really a coalition environment."[28] The spark teams did what they could to improve communications between the countries providing aid.

Throughout the operation, even simple communications between national partners remained a major problem and were inevitably handled in informal ways. For example, to get a helicopter to deliver ten tons of hospital equipment to a German field hospital near Teunam, about eighty miles southeast of Banda Aceh, a German Red Cross worker, Ina Bluemel, simply walked up to an American pilot "asking for a helicopter."[29] The pilot agreed and delivered the supplies. This was perhaps the most typical form of coalition communications during the operation.

Another problem with communications concerned bandwidth allocation. Much of the equipment was not designed to allow changing the bandwidth, with the result that far too much was permanently allocated to equipment that was unused. The U.S. Navy allocated approximately half of all of its bandwidth for telephones, but the average usage was low. It would have been much more efficient if bandwidth could have been reallocated as needed: "Don't need to buy more satellites, just need to allocate bandwidth where it is needed. Connectivity needs to be given top priority."[30]

The Impact of the Media and Global Communications

Another side of the communications question was the transmission and reception of outside information, particularly to and from the foreign press, over the Internet, and with e-mail. In general, this worked out in a positive way. From the beginning, the U.S. Navy's skill in conducting humanitarian relief operations was favorably commented on by the international press. *Abraham Lincoln* also served as an important base of operations for international film crews and newsmen, including, as noted, *60 Minutes*. While the film content of CNN reports was generally gruesome, showing dead bodies and the general devastation, in the midst of it a news clip would show U.S. helicopters delivering aid. In other words, when *Lincoln* arrived, America was there.

To assist in filing video reports, the ship used fast file transfers over the SIPRNET, since this was least restricted for those with the right clearance. Although this was expensive equipment and was not normally used so regularly, it proved to be the best way of getting video out quickly. Still, there were notable backlogs. Lieutenant Commander John Daniels later recalled that because of the enormous amount of information being transmitted, "stacks of email and phone were never answered because traffic and requests were just enormous. Tremendous push to get video out."[31]

Captain Bob Aronson recalled the effect on morale of the first CNN reports on UNIFIED ASSISTANCE: it was "a great boost to see this report on TV after the horrific sights and smells they encountered the first day they arrived."[32] Most of the ships were able to receive satellite TV and so could watch the media reports on television within minutes of seeing the reports being filmed; this provided an immediate sense of accomplishment among the ships' crews and shore parties. Access to e-mail, to keep in touch with family as well as colleagues not assigned to UNIFIED ASSISTANCE, meant that crew members constantly received new information about the humanitarian relief mission. This positive feedback, in turn, motivated them to work even harder to assist victims in the devastated areas.

Happy faces communicating refugees' gratitude.

There was a potential downside, however, in the fact that crew members could write personal e-mails and even Weblogs from the ship. Media reports quoted from such sources private criticism of the various groups of aid workers, one "blog" denouncing them as a traveling circus who wasted valuable helicopter time being ferried to shore and then back to their guest quarters.[33] Such opinions, while representing a minority view, tended to undermine the image, carefully crafted by the PAOs, of unconstrained cooperation between U.S. Navy personnel, media, and civilian aid workers.

Vertical and Horizontal Communications Challenges

Due to the particular joint task force structure adopted during the first days of the operation—Utapao, not the Carrier Strike Group 9 staff or USS *Abraham Lincoln,* was designated as the headquarters—communications between Utapao and *Abraham Lincoln* time and time again proved to be a stumbling block. When the carrier arrived off Indonesia on 1 January 2005, it became the main communications hub linking Utapao and Honolulu, all the major naval commands, and the group's ships and their teams working ashore. The early establishment of this communications network proved particularly important, saving much time and allowing humanitarian relief efforts to be carried out with greater efficiency.

But it took a long time for Utapao to acquire global communications, so the advance team was "walking into a black hole themselves." In addition to hooking up unclassified but sensitive Non-secure Internet Protocol Router Network (NIPRNET) and classified Secret Internet Protocol Router Network (SIPRNET) communications, dozens of commercial cell phones were purchased from local Thai companies: "Local Thai cellular phones are the convenient way to go," argued Marine colonel Medio Monti, who installed the communications system. Utapao also constructed a website, which allowed other militaries and aid groups to monitor the U.S. military's activities. By making information available through the website, the planners at Utapao hoped to avoid "duplication and waste." Lieutenant General Blackman especially encouraged "his deputies to consult other nations regularly to avoid overlap."[34]

One problem proved to be moving communications equipment to Utapao. The high-speed, shallow-draft catamaran *WestPac Express* left Okinawa on 4 January for Thailand carrying 630 tons of equipment, including communications gear. After a 2,300-mile journey, *WestPac Express* arrived in Chuksamet, Thailand, on the 10th.[35] Along with thirty-five vehicles, including seven-ton trucks, "Hummers," and forklifts, it unloaded thirty pieces of communications equipment and thirty Marines of 7th Communications Battalion, III Marine Expeditionary Force Headquarters Group.[36]

In light of the fact that U.S. humanitarian operations in Thailand were wrapped up less than two weeks later, on 22 January, this was a clear example of too little, too late. By contrast, the sea-based communications network worked well; *Abraham Lincoln* arrived on station with unified communications, command, and control already up and running. Otherwise, structural differences interfered with rapid communications both vertically and horizontally. While there were hundreds of people working on organizing the mission in other places, there were only a handful of people at the "point of the spear."[37] As one critic was quick to point out: "Not much value [was] added by hundreds of people at PACOM, while only 40 people on the ground were doing all the real work. JTF structure might not be right."[38]

Many other problems occurred when chat rooms used different software. Some used Asyncrony and Vogue, but they were "extremely painful" in a naval context, since they involved Web-based messaging. Problems appeared with accessing chat rooms; also, any message over a megabyte in size was too big to transmit. Some groups, including people on shore or on *Mercy*, tried to use Webex. Unfortunately, this system was generally not used on ships because it required a 3.1-megabyte applet download. Once downloaded it was "found that the chat room was password protected and the only person . . . who knew the password was on leave, so they could never get in the room to use it."[39]

The international language—"thumbs up!" AD3 Jason Shireman (HC-11 Det 2), at Lamno, Aceh Province.

As for horizontal communications, some were surprised how long it took to get mobile communications into the theater, forcing shore parties to use cell phones and telephones. Shore establishment communications was particularly important for ensuring clarity: "For every layer of command and control that information went through, it became distorted or changed in one way, shape, or form." There was a need to "streamline that whole hierarchy." One solution to this confusing communications picture would have been to eliminate unnecessary nodes and "flatten" the hierarchy. This would have allowed *Abraham Lincoln* to communicate directly with Pacific Command: "E-mail would have worked better if they could have sent it straight to PACOM, rather than through the CSF as intermediary. This slowed the information flow. Also, Internet Relay Chat was being used like radios, but without the discipline inherent in traditional radio communications."[40]

* * * * * * *

Throughout the operation, the inherent disconnects caused by *Abraham Lincoln*'s reliance on SIPRNET and Utapao's reliance mainly on NIPRNET caused notable communication problems. Even when Utapao acquired full access to the unclassified net it was sparsely utilized. Only on 15 January was CSG-Indonesia in Medan up and running on SIPRNET, and on 18 January CSG-Indonesia Forward in Banda Aceh acquired both SIPRNET and NIPRNET. Almost by default, this meant that the forces afloat could often communicate more quickly and easily with Pacific Command than within the theater of operations, which undermined the vertical and horizontal command structure. In hindsight, Admiral Doran admitted that it might have been easier to make *Abraham Lincoln* the operation headquarters.[41]

Miscommunication and rumor were still rampant. One incident that exemplifies this problem involved a large box of dried noodles that accidentally fell from a helicopter onto the tarmac at Banda Aceh airport. No one was injured, but as this story went through the rumor mill it expanded into a pallet crushing several people. The immediate effect was that aid workers no longer wanted to fly in the helicopters, and it took several days for everything to return to normal. This problem was certainly linked to insufficient horizontal communications, not always dependable even when available. Thus, a seemingly trivial story could spread and become exaggerated beyond all proportion. The spread of other such malicious rumors was to affect the medical mission as well.

Notes

1. Epigraph from Rear Admiral Douglas Crowder, CCSG-9, Naval Historical Center [hereafter NHC] Oral History, 20 February 2005.
2. Kathleen T. Rhem, "'Conditions for Transition' Met in Thailand, Sri Lanka," American Forces Press Service, 15 January 2005.
3. Commander Dan Verheul, CCSG-9 staff N2/ ACOS Intelligence, NHC Oral History, 20 February 2005.
4. Navy Warfare Development Command, "Humanitarian Assistance/Disaster Relief (HA/DR) Operations Planning," TACMEMO 3-07.6-05 (Newport, R.I.: August 2005), annex A.
5. Peter Boylan, "Military to Continue Tsunami Aid," *Honolulu Advertiser*, 11 January 2005.
6. Captain Bob Aronson, CCSG-9 staff N6/ ACOS C41, NHC Oral History, 15 February 2005.
7. Commander Vince Quidachay, CCSG-9 staff N4, NHC Oral History, 20 February 2005.
8. Lieutenant Christine McManus, DesRon 9 staff air operations officer, NHC Oral History, 15 February 2005.
9. Commander Ted Williams, executive officer, VAQ-131, NHC Oral History, 20 February 2005.
10. Commander Dan Boyles, executive officer, HSL-47, NHC Oral History, 20 February 2005.
11. Lieutenant Commander Jason Carter,CCSG-9 staff communications officer, NHC Oral History, 20 February 2005.

12. AO1 Kevin Parra, EODMU, 11 Group, NHC Oral History, 20 February 2005.

13. Carter oral history.

14. Admiral Walter Doran (Ret.), interview with Elleman, Arlington, Virginia, 23 October 2006.

15. Aronson oral history.

16. Sunoy Banerjee, PhD, CNA Representative, USS *Abraham Lincoln,* NHC Oral History, 20 February 2005.

17. Carter oral history.

18. Parra oral history.

19. Captain Matthew Klunder, DCAG, CVW-2, NHC Oral History, 20 February 2005.

20. Commander Chris Meyer, chief engineer, USS *Abraham Lincoln* (CVN 72) NHC Oral History, 20 February 2005.

21. Aronson oral history.

22. Banerjee oral history.

23. Carter oral history.

24. Banerjee oral history.

25. Ralph A. Cossa, president of Pacific Forum CSIS, "U.S. Military Provides 'Logistical Backbone' for Tsunami Relief" *eJournal USA: Foreign Policy Agenda* 9, no. 3 (November 2004), usinfo.state.gov/gi/Archive/2005/Mar/05-433015.html.

26. Mark S. Leavitt, Jeffrey M. Vorce, and Michael M. Hsu, "For Compassion and Country: UNIFIED ASSISTANCE," U.S. Naval Institute *Proceedings* 131, no. 4 (April 2005).

27. Captain Brian Roby, CCSG-9 chief of staff, NHC Oral History, 20 February 2005.

28. Banerjee oral history.

29. Denis D. Gray, "US Military Plans on Long Stay to Airlift Supplies to Coast," *Boston Globe,* 10 January 2005.

30. Banerjee oral history.

31. Lieutenant Commander John Daniels, PAO, USS *Abraham Lincoln* (CVN 72), NHC Oral History, 20 February 2005.

32. Aronson oral history.

33. Philip Sherwell and Inigo Gilmore, "U.S. Navy Officer Attacks 'Travelling Circus of Aid Workers' for Impeding the Tsunami Relief Effort in Indonesia," *London Sunday Telegraph,* 30 January 2005.

34. James Brooke, "Ad-hoc Planning Goes into Marine Aid Effort," *New York Times,* 6 January 2005.

35. Samantha L. Quigley, "High-Speed Vessel Proves Worth during Disaster Relief Mission," American Forces Press Service, 28 January 2005.

36. Sergeant Mandy L. McCammon, "WestPac Express Supports Humanitarian Assistance Mission," Hornblower Marine Services, 10 January 2005.

37. Ibid.

38. Banerjee oral history.

39. Ibid.

40. Ibid.

41. Doran interview.

CHAPTER SEVEN

The Medical Mission

USNS *Mercy* Signals Political Will

This was the biggest natural disaster in the history of Indonesia, bigger even than Krakatoa. . . . There are those who compare us with Thailand and how quickly they acted. The scale and magnitude are different. The local government collapsed, the infrastructure was destroyed, there was no electricity, fuel, communication or transportation. We mobilized what resources we had but needed help from outside nations, and I give them our greatest thanks.

SUSILO BAMBANG YUDHOYONO, PRESIDENT OF INDONESIA

17 JANUARY 2005

The USNS *Mercy,* one of two U.S. hospital ships in the Military Sealift Command, was activated by presidential order on 1 January 2005.[1] When its commanding medical officer, Captain David Llewellyn, was informed during late December that *Mercy* might be activated, he and his crew began logistics and medical planning: "Both were a challenge." All this was accomplished in record time, however, and the ship was soon ready to embark on its humanitarian mission: "Getting underway for a six-month deployment in less than a week was both a monumental challenge and a monumental accomplishment."[2]

By 8 January *Mercy* had loaded hundreds of active-duty and reserve physicians, nurses, technicians, and other military and civilian medical workers and was on its way from San Diego to Indonesia. On board the ship were twelve operating theaters, magnetic resonance imaging (MRI) equipment, and four seventy-five-thousand-gallon-per-day evaporators with "1.2 million gallon holding capacity."[3] The ship and medical staff eventually treated more than 9,500 patients in humanitarian missions in Indonesia, East Timor, and Papua New Guinea before returning to San Diego on 8 June 2005. This story was especially remarkable because the *Mercy* had not been deployed for thirteen years, not since DESERT STORM, and yet it encountered no significant engineering problems during the entire deployment.[4]

The hospital ship *Mercy* became an important symbol of how the United States could help countries in distress. Admiral Thomas Fargo saw *Mercy* as not just a medical asset but a golden opportunity to reengage Indonesia politically. He had lobbied DoD to activate *Mercy* to improve U.S. and Indonesian relations. Given the time it took to arrive in Indonesia, Fargo perceived, the primary mission of *Mercy* was not just to provide emergency care but also, and even more importantly, to help guard against epidemics by reconstituting the "Indonesian medical infrastructure."[5]

Political Considerations

Captain Llewellyn, head of *Mercy's* medical treatment facility, who was visiting his father in San Francisco during Christmas vacation, later recalled that there had been no news of *Mercy's* deployment until he received a call in late December from the Military Sealift Command and another from the Third Fleet surgeon asking him "to send a PowerPoint presentation to PACFLT [the staff of Commander, Pacific Fleet] explaining the capabilities of USNS *Mercy*."[6] The crew and medical staff of *Mercy* began preparing on 31 December 2004 for deployment.

Opinions were mixed as to how much use the ship could be, considering how long it would take to reach Indonesia. The decision to deploy came "directly from the President"; Captain Tim McCully was to observe that he had "never in his career seen any deployment that had this level of interest, from CNO's [Chief of Naval Operations] office on down." Even when the ship was ready to leave, there were "discussions even at that time at the 4-star level deciding whether to get the ship underway."[7] *Mercy* is a "deployable 'national asset.' As such, only the highest authority can authorize its deployment. Once deployed, the world is put on notice especially in time of war (e.g., DESERT STORM)." On the day the ship left port, Captain Smith talked with three different people on the phone: "One told him to get underway. The other told him not to get underway. The third person did not know but said he would find out." Nobody in the DoD wanted *Mercy* to get under way only to turn around and "come back to San Diego a few hours or a few days later." When *Mercy* did depart from San Diego, it sent thereby a strong message that the United States was fully "committed to tsunami relief efforts in South Asia."[8] Once on station *Mercy* quickly proved both its medical and political worth.

Captain Nathan Smith, commanding officer of the USNS *Mercy*, credited "Fargo and his vision for making deployment happen."[9] Admiral Walt Doran agreed, recalling that once Fargo made up his mind, he "never wavered."[10] In addition to the clear political benefits of deploying *Mercy*, had there been large-scale epidemics in Indonesia, "having that resource in theater would have been hugely valuable." Also, *Abraham Lincoln* would eventually redeploy, and *Mercy* would help fill the "vacuum when the carrier steamed out of the area."[11]

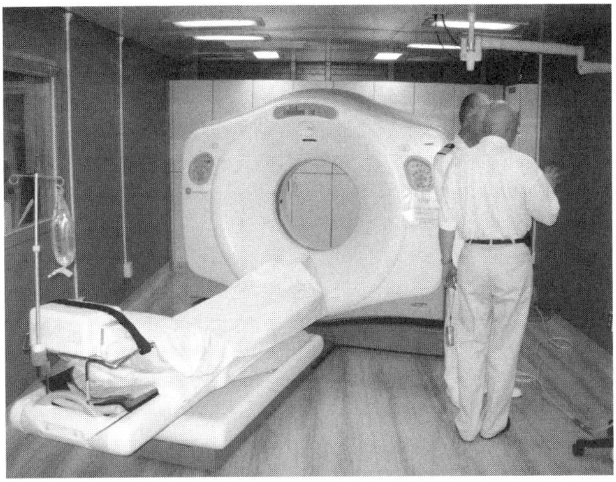

Magnetic resonance imaging (MRI) on the USNS *Mercy*, en route to Aceh Province.

The Situation in Indonesia

The Indonesian government apparently either did not know that *Mercy* was coming or simply could not decide what to do with it. As a result, when *Mercy* arrived, the ship and its crew were not permitted to go to work immediately;[12] a five-day delay ensued, while the ship sat off the coast of Banda Aceh.[13] An advance party led by Admiral Bill McDaniel (Ret.) had arrived in Banda Aceh ahead of the ship to identify potential patients to send back to the ship for medical care.[14] According to Captain Timothy Bemiller, however, from the very beginning it was unclear exactly what *Mercy* would be doing. As a result, "the mission evolved and so did our planning. We had to be flexible. . . . [W]e never had clarity on what the mission was."[15]

Mercy was eventually sent to northern Sumatra, and Commander Marquez Campbell, a medical planner, was one of the first to go ashore, inspecting the hospital, airport, and potential landing zones. He would recall, "Seeing the damage and destruction from a helicopter was staggering." On board, the Operations Center was run like a "battle watch," as helicopters took volunteers from the ship to shore and brought patients back to the ship.[16]

Unfortunately, the time it took to reach Aceh Province and then delays in getting started meant that by then many seriously ill or injured patients had already died. In fact, even in early January it was already clear that the "likelihood of saving sick patients still stranded in remote areas was fading." As one Australian doctor, Alan Garner, explained, "Even some of those who arrive now would have a tough time surviving. A 20-year-old guy brought in today said he was the sole survivor of his village. He was washed out to sea, walked for two days and was picked up by the American helicopter. But I suspect he's going to die, and he will be the last of his village."[17] One dramatic success story, however, was an eleven-year-old boy named Iqbal who was found clinging to a piece of driftwood over a mile out to sea. All the rest of his family had been killed. He was brought to *Mercy* by helicopter and treated for a respiratory illness known as "tsunami lung," caused by swallowing bacteria-laden water. A week later Iqbal was on the road to recovery, and within four weeks he was able to leave the ship.[18]

Since *Mercy* arrived many weeks after the tsunamis, "very little was used" of its huge capacity.[19] Captain Llewellyn later stated that due to delays "there were not very many acute cases to treat. Instead, most medical cases were nonacute but necessary and important. Because of the widespread destruction of local hospitals, *Mercy* treated mostly chronic diseases, and also head and neck tumors."[20] That is, what *Mercy* mainly treated were not emergency cases linked directly to the tsunamis but more long-term health concerns. The head of radiology, Lieutenant Commander Stephen Ferrera, described the first day on station after being cleared by the Indonesian authorities to treat patients: "Did operation on person with appendicitis. More patients started arriving quickly. Things went well as they were brought onboard by helicopter. Working with other nationalities was at first stressful but all adapted. Civilians and organizational standpoint also caused some discomfort but all adjusted. Good PR for Navy medicine."[21] *Mercy,* which has superior laboratories, conducted over five thousand laboratory procedures during a six-week period from early February to mid-March 2005.[22]

Helicopter operations became the crucial lifeline between *Mercy* and the shore, and during the transit repairs were made to the flight deck.[23] Captain Llewellyn had to coordinate constantly with the other ships, since the helicopters his ship would fly actually belonged to the USS *Essex* and *Fort McHenry.*[24] Helicopters in fact made the sea-based medical mission possible, flying over 1,100 helicopter sorties during *Mercy*'s deployment: "The helicopter ride to and from the ship was around 15 to 20 minutes long. . . . If a stretcher and patient were onboard, the stretcher took 3 seats so only 8

The Military Sealift Command (MSC) hospital ship USNS *Mercy* off Aceh Province.

persons could be transported."[25] The ship also was visited by many other American and foreign helicopters, including a CH-53 from *Essex*, an H-3 from Germany, a Huey from the Spanish Navy, and an SH-60 from the Japanese Maritime Self-Defense Force.[26]

Mercy and Project Hope

Admiral Fargo thought coordinating with private relief organizations should be an important goal of UNIFIED ASSISTANCE, since nongovernmental organizations could use *Mercy* as a base of operations: "This may be an opportunity to use *Mercy* in a very creative way."[27] Coordination between *Mercy* and Project Hope had first been suggested several years before at the senior government and military level. When he heard about the Southeast Asian tsunami, Admiral James Lyons (Ret.) immediately called the Chief of Naval Operations, Vernon Clark, and urged him to get in touch with John Howe, director of Project Hope.[28] He did so, to discuss this "novel idea"—UNIFIED ASSISTANCE would be the first Project Hope–U.S. Navy mission.[29] Admiral Clark then met with Secretary of Defense Donald Rumsfeld to convince him of the importance of this civil-military program; Rumsfeld told him to "go for it."[30]

Following intense discussions, Project Hope, based in Millwood, Virginia, agreed in a memorandum of understanding to locate and select medical volunteers to serve on board *Mercy*.

Project Hope was able to pick from some of the best doctors and nurses in the country. The word went out to health professionals across the nation that *Mercy* would be dispatched to care for survivors of the tsunami and that volunteers were needed to help staff the ship in one-month rotations. Project Hope was inundated with applications; within a matter of days, it had received over four thousand applications to fill the 210 openings for doctors and nurses.[31] For the first of three planned rotations, ninety-three highly qualified professionals from dozens of the country's leading centers of medicine and education were picked.[32] Massachusetts General Hospital, in Boston, Massachusetts, alone sent "30–40 people."[33] Project Hope leaders General Harold Timboe (Ret.), former head of Walter Reed Hospital, and General Bill Bester (Ret.), former head of the Army Nurse Corps, boarded *Mercy* at Honolulu. *Mercy* received its first volunteer group of sixty nurses and thirty doctors at Singapore.[34]

This cooperative civil-military program was of great help to the U.S. Navy, since embarking sufficient naval medical personnel in San Diego was a major problem. Pulling staff from the local Naval Medical Center would have been very expensive and could have left it with insufficient staff to carry out normal duties. So allowing Project Hope volunteers to come onboard in their place was a cost-saving measure and also enabled the Naval Medical Center at San Diego to avoid service interruption.[35]

Due to shortness of time, Project Hope volunteers conducted short, one- or two-day orientation tours on board a similar hospital ship, USNS *Comfort,* stationed in Baltimore, Maryland. Not having completed a lengthy shipboard orientation, many had to undergo an additional orientation upon arrival on *Mercy* itself.[36] Later in the deployment, when new Project Hope volunteers replaced the first set, there "was no time for any [orientation-related] training and teambuilding." Commander Jean Comlish was in charge of keeping track of the Project Hope volunteers. Their numbers constantly changed, but eventually volunteers were assigned primarily to three areas: the "Emergency Room, Intensive Care, and the Wards."[37]

The numerous Project Hope volunteers meant that "the ship had to work out berthing and other issues related to the additional crew members." Yeoman 2nd Class Daniel Konzek later recalled that "his department was tasked to develop a process to track crew members on the beach and on the ship" as well as to "track the patients who came on board."[38] After a month, when a new Project Hope team arrived, everything started all over again with new training.[39]

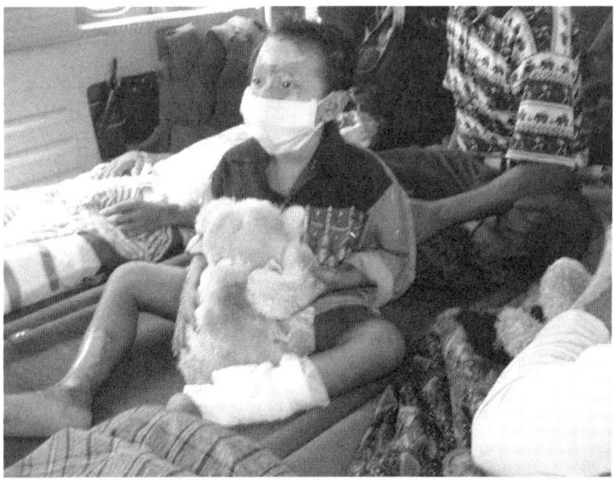

Injured child being airlifted to medical facility, Aceh Province.

At first, the relationship between the military and volunteers was described as "awkward," because the volunteers knew so little about the military. For example, the volunteers would not show up for muster on time. After a while, however, this all changed, and "we got comfortable with them and they got comfortable with us."[40] The naval personnel were "very impressed with [the volunteers'] professionalism."[41] According to Commander Kurt Hummeldorf, "the benefits of HA/DR missions are many and are well worth the cost. . . . Navy medicine combined with NGO support is a powerful tool to win the hearts and minds of people around the world."[42]

Long-Term Medical Impact

In addition to providing emergency medical assistance, *Mercy* provided Indonesia longer-term medical benefits: "Once on station, we made oxygen and filled tanks. We had laundry capabilities. We had electricians who could fix equipment. There were many other examples as well. And of course, we primarily provided medical HA/DR."[43] In addition, the ship's optometrists saw about two hundred patients a day, and the ophthalmologist performed cataract surgeries on board.[44]

Mercy would ask, "How can we help?"; this "approach endeared us to the hospital staff at Banda Aceh and to the Indonesian people in general."[45] One way to help was medical training at the hospital at Banda Aceh. It soon became clear that many of its personnel had been killed by the tsunami; foreign NGOs were manning the hospital. Instead of just providing medical care, therefore, *Mercy* also assumed responsibility for training. So many Indonesian medical instructors had perished in the tsunami that the American doctors "filled the instructor void while we were in Banda Aceh." Lieutenant Commander Suzanne Clark took over medical training: "While there, she looked at the education and training opportunities that we could provide to the Indonesian medical staff. When we arrived, the Australians were conducting training. After the Australians left, we continued the training and education that they had been conducting."[46] She taught for an average of three hours a day.

One limiting factor was translators. Most translators worked around the clock. Many of them were medical personnel from Indonesia, and probably from ten to twenty were on board at any one time, living in the enlisted berthing spaces.[47] Before there were sufficient translators, "many of the patients who came onboard were scared. [But] after their surgery and post-operation care, the patients cried and hugged us when they left the ship." The friendship was even deeper with the local translators, without whose help little or nothing could have been done.[48]

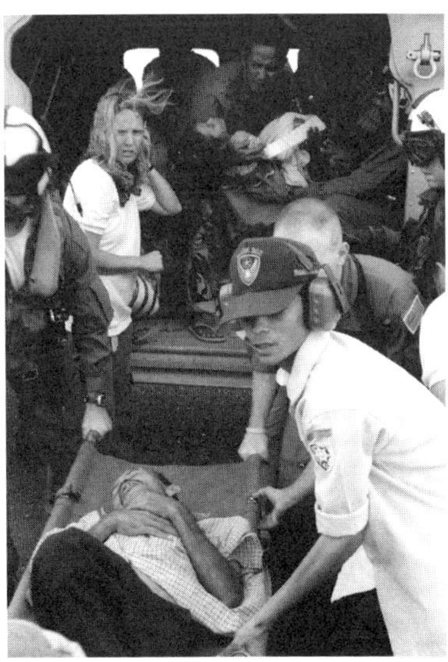

Medical evacuation personnel offload patients at Banda Aceh airport.

Due to force protection constraints, none of the *Mercy* personnel could stay overnight in Banda Aceh; everyone had to return to the ship at the end of each day. There was also great concern that no Indonesian patients die aboard *Mercy*, for fear that the ship and crew might be blamed for the death—there might have been malicious rumors, perhaps official accusations, that the foreign doctors had killed them. However false, they might have sparked anti-American feeling or even riots. Therefore, as a matter of policy, terminally ill patients were returned home so that they could die with their families.[49]

The *Mercy* mission was scheduled to last for ninety days. According to Commander Karen McDonald, "the Indonesian government had given us a set date to leave Banda Aceh. We had to adjust our patient flow and treatment to meet that deadline. Commander McDonald stated that it was hard when mothers would beg you to take their child to the ship, and you could not.[50] *Mercy* left Indonesian waters on schedule and continued on to its next mission, in East Timor.

* * * * * * *

Sending *Mercy* to Indonesia was clearly a humanitarian success. When Admiral Fargo briefed him in mid-January 2005, President Bush was "tremendously pleased."[51] But Captain Smith's own assessment was that it was little short of a miracle that nothing bad happened: "The analogy is taking a twenty-year-old vehicle, re-build the engine, and drive the vehicle from San Diego to New York City and back. Luckily, we didn't lose any medical equipment." As it happened, however, the maintenance that had been performed during the ship's years of inactivity by the Reduced Operation Status crew and the Military Sealift Command's project manager proved to have been just right, and "the ship held up well."[52] As a result, *Mercy* became a potent symbol of American assistance to Indonesia.

As throughout UNIFIED ASSISTANCE, communications was key to success. According to Lieutenant Commander Erik Threet, communications at the beginning of their stay in Aceh Province were problematic, but "after awhile, we worked out all the issues and communications were fine for the remainder of the Banda Aceh mission." If a patient needed to be taken to the ship, "we could communicate the patient's condition while the patient was in-transit to the ship. Once the patient arrived at the ship, the ship's medical crew knew what the patient's condition was. This helped save time and lives. Communications were paramount."[53]

According to Captain Llewellyn, the Indonesian deployment established what has since been called the "*Mercy* model," in which the mind-set of the ship's crew was to ask, "How can we help[?] . . . How can we contribute and help the recovery process?" The first lady, Laura Bush, sent a letter to Project Hope and *Mercy,* telling them that they represented the "compassionate heart of America." Llewellyn reiterated that it was "hard to imagine what it was like to have endured such a disaster and how gratifying it was to help."[54]

Notes

1. Epigraph quoted in "We Are Very Grateful" (interview with President Susilo Bambang Yudhoyono), *Time Asia*, 17 January 2005, available at www.time.com/time/asia/covers/501050124/interview.html.
2. Captain David Llewellyn, commanding medical officer, USNS *Mercy* (T-AH 19), Naval Historical Center [hereafter NHC], Oral History, 2 June 2005.
3. Captain Tim McCulley, MSC Liaison, PACFLT Staff, Pearl Harbor, Hawaii, NHC Oral History, 2 March 2005.
4. Captain Tim McCully, USNS *Mercy* (T-AH 19), NHC Oral History, 7 June 2005.
5. Captain Nathan Smith, USNS *Mercy* (T-AH 19), NHC Oral History, 6 June 2005.
6. Llewellyn oral history.
7. McCully oral history.
8. Smith oral history.
9. Ibid.
10. Admiral Walter Doran (Ret.), interview with Elleman, Arlington, Virginia, 23 October 2006.
11. Admiral Thomas Fargo, telephone interview with Elleman, 30 July 2006.
12. Llewellyn oral history.
13. Commander Jean Comlish, USNS *Mercy* (T-AH 19), NHC Oral History, 3 June 2005.
14. Commander Kurt Hummeldorf, USNS *Mercy* (T-AH 19), NHC Oral History, 3 June 2005.
15. Captain Timothy Bemiller, USNS *Mercy* (T-AH 19), NHC Oral History, 7 June 2005.
16. Commander Marquez Campbell, USNS *Mercy* (T-AH 19), NHC Oral History, 7 June 2005.
17. Jane Perlez, "Pace of Global Aid Effort Picks Up on Indonesian Coast," *New York Times*, 4 January 2005.
18. Isadore Rosenfeld, "A New Way to Treat the World," *Washington Post-Parade*, 12 June 2005, p. 5.
19. McCulley oral history.
20. Llewellyn oral history.
21. Lieutenant Commander Stephen Ferrera, head of radiology, USNS *Mercy* (T-AH 19), NHC Oral History, 5 June 2005.
22. Lieutenant Commander Jeosalina Serbas, USNS *Mercy* (T-AH 19), NHC Oral History, 5 June 2005.
23. ABFC Windsor Solar, USNS *Mercy* (T-AH 19), NHC Oral History, 5 June 2005.
24. Ferrera oral history.
25. MA1 Scot Valle, USNS *Mercy* (T-AH 19), NHC Oral History, 6 June 2005.
26. Solar oral history.
27. Bradley Graham, "U.S. Relief Effort in Asia Not Affecting Iraq Forces," *Washington Post*, 5 January 2005.
28. Admiral James Lyons, telephone interview with Elleman, 20 October 2006.
29. John Howe, telephone interview with Elleman, 18 October 2006.
30. Admiral Vernon Clark (Ret.), telephone interview with Elleman. 17 October 2006.
31. Howe interview.
32. "Project HOPE Sends 93 Medical Volunteers on a Humanitarian Aid Effort on the USNS *Mercy*," 4 February 2005, www.projecthope.org/news/020405.htm.
33. Ferrera oral history.
34. Llewellyn oral history.
35. Smith oral history.
36. Lieutenant Commander Suzanne Clark, USNS *Mercy* (T-AH 19), NHC Oral History, 6 June 2005.
37. Comlish oral history.
38. YN2 Daniel Konzek, USNS *Mercy* (T-AH 19), NHC Oral History, 4 June 2005.
39. Llewellyn oral history
40. HM2 Jason Moore, USNS *Mercy* (T-AH 19), NHC Oral History, 5 June 2005.
41. Ferrera oral history.
42. Hummeldorf oral history.
43. Ibid.
44. Lieutenant Robert Senko, USNS *Mercy* (T-AH 19), NHC Oral History, 7 June 2005.
45. Bemiller oral history.
46. Clark oral history.
47. Ferrera oral history.
48. Moore oral history.

49. Llewellyn oral history.

50. Commander Karen McDonald, USNS *Mercy* (T-AH 19), NHC Oral History, 6 June 2005.

51. Fargo interview.

52. Smith oral history.

53. Lieutenant Commander Erik Threet, ITC Bobbi Jo Williams, IT1 Joshua Davis, IT1 John Opsteegh, and IT1 Damion Jasso, NHC Oral History, USNS *Mercy* (T-AH 19), 6 June 2005.

54. Llewellyn oral history.

CHAPTER EIGHT

Wrapping Up the Mission
Lessons Learned and Recommendations

This is the largest natural disaster of my lifetime and it changes the rule set. It can't help but be positive, in my view.

REAR ADMIRAL DOUGLAS CROWDER, *LINCOLN* STRIKE GROUP COMMANDER

2 JANUARY 2005

After less than six weeks, the U.S. military forces accomplished their primary mission.[1] America's humanitarian response to the tsunami crisis had been unprecedented. In the words of Andrew Natsios, Agency for International Development (USAID) administrator, "The response of the American private sector has been extraordinary. I am told that in terms of dollars pledged or donated, this is the second-largest disaster response in American history, second only to 9/11. America responded to our own tragedy and we are equally concerned about tragedies elsewhere."[2]

The Indonesian government appreciated U.S. aid, but it asked that all American military forces be withdrawn by March 2005. Some Indonesians were concerned at an early withdrawal, but Rear Admiral Crowder disagreed: "The bottom line is: I don't share that same concern. We're reaching a point where there's going to be a transition to sustain relief and not an acute emergency got-to-have-it-now relief that we saw in the first couple of weeks."[3]

Following the departure of *Abraham Lincoln* and its support ships on 10 February 2005, CSG-Indonesia was shut down, followed by the Combined Support Force 536 headquarters on 12 February 2005. JTF 536 was set up again to serve temporarily in its place; finally, it too was closed down, on 23 February 2005, and UNIFIED ASSISTANCE officially ended. The operation lasted less than two months from beginning to end, but many valuable lessons had been learned, in particular the importance of tsunami early warning systems and accurate damage reporting, rapid and flexible sea basing, robust logistical trains, helicopter access, efficient communications, and world-class medical care long after a disaster.

Warning and Reporting

An Indian ocean tsunami early warning system could have saved many lives. After the tsunamis hit the biggest problem was inaccurate reporting; local governments and the international press at first badly underreported the number of dead and missing. Had a more accurate picture been immediately available, the U.S. Navy's arrival, rapid as it was, could have been accelerated by a full day if not more, which would have made a huge difference to many tsunami victims.

The poor reporting was not strictly the fault of the media, which were constrained to report what reporters themselves either saw or were told by local officials. The almost total destruction of local communications infrastructure meant that few reports were

Early tsunami warning systems could have saved many lives.

ever received from the hardest-hit areas. Still, more accurate numbers provided earlier by the press could have made the international response both faster and more focused on areas that needed help the most.

The U.S. Navy's decision to send over two dozen large and capable ships, including the carrier *Abraham Lincoln*, was made to allow the response to ramp up as the scope of the disaster became clear. This was not an accident or afterthought; the possibility of escalation was factored into the decision-making process from the very beginning.[4] This made possible the rapid arrival of highly flexible sea-base ships to coordinate the humanitarian relief mission.

Rapid and Flexible Sea Basing

Sea basing merged capability with cultural awareness and proved to be particularly important because the destruction throughout Aceh Province was so widespread; it was later to serve as a model for the post-Katrina cleanup on the Gulf of Mexico.[5] One participant who had contributed to damage control after 9/11 compared that response with the tsunami relief mission: "Initial reaction on going ashore . . . flying over the area and seeing the devastation was mind-boggling."[6]

Abraham Lincoln provided a unique capability, one that is "unmatched anywhere in the world" and gave the United States the greatest possible range of operational choices and capabilities.[7] Some critics, pointing to the six-million-dollar-a-day price tag, argued that an aircraft carrier was "not a particularly efficient vehicle for delivering aid and succor to scattered, ravaged villages."[8] High-speed ships like *WestPac Express,* and perhaps even the new Littoral Combat Ships (LCS), could have arrived at the disaster area in far less time and operated at a far lower daily cost.

Or, as General Mike Hagee, Commandant of the U.S. Marine Corps, later put it, "What are your sea connectors and how fast can they go? Can we take some of the ships on the drawing boards, which are quite fast, relatively small, maybe don't have all the protection that a gray-bottom would have, and could we use that as part of the sea basing?"[9] Depending on the nature of the crisis, such vessels could have been quickly reconfigured for roll-on/roll-off to respond to specific humanitarian missions.

But such criticism generally ignores the enormous psychological impact of sending an aircraft carrier. As Admiral Fargo emphasized, "You have already paid for the fixed costs, like fuel for the nuclear reactor. . . . Nobody sat around thinking that we will do this if there is a supplemental."[10] While many other countries assisted under the auspices of the United Nations, and in doing so exerted "soft power," their contributions were limited in comparison to the U.S. Navy's hard-power assets.

Sea basing brought essential humanitarian supplies to Aceh Province.

The carrier was much more than just a delivery system but the focus of virtually all decision making, logistics coordination, communications, and support for a myriad of diverse activities: "Sea-basing was required, not really delivery from the sea."[11] For many victims the American "carrier projected not only power but hope"; in particular, the "UN people were appreciative of what the 'big grey house' provided."[12]

According to Rear Admiral Crowder, sea basing also allowed the U.S. Navy to "low-key our presence," reduce the "footprint," and thereby minimize cultural friction.[13] Sea basing provided by *Abraham Lincoln* was "less intrusive"—"such missions are more likely to be welcomed by nations concerned with the presence of United States military personnel 'occupying' bases ashore."[14]

Finally, sea basing combined well with force protection, since the ability to return all personnel each night from Banda Aceh to their ships helped prevent "the relief mission from becoming a potential target for terrorists."[15] During UNIFIED ASSISTANCE only about 2,500 American troops, about 15 percent of the total force, were based on land throughout the entire region.[16] According to Admiral Fargo, "The sea base means you don't have the force protection, and you solve a lot of political issues. . . . These are sovereign nations who have a great deal of pride. And the fact that you don't have a great number of people on the ground 24 hours a day is a great advantage."[17]

The U.S. Navy's rapid arrival, long-term presence, small "footprint" on the land, and sensitivity to the local culture showed Indonesians that Americans cared about them as individuals. In this way, "*Abraham Lincoln* was the face of America."[18]

Creating a Robust Logistical Train

Sea basing could coordinate and funnel essential humanitarian supplies. While the total tonnage of supplies delivered was one measure of the operation's success, the coordination required to deliver it to the right locations was even more important: "Sea-based relief efforts capitalize on the efficiencies gained from the existing support infrastructure of ships (communications, food, shelter, medical facilities, and fuel)."[19]

With regard to the size of the logistical flow, UNIFIED ASSISTANCE in some ways rivaled a Berlin Airlift at sea, conducted over forty days rather than four hundred. After only two weeks, military logistics had moved sixteen million pounds of relief supplies.[20] According to Admiral Fargo, from late December through mid-February the U.S. military delivered over twenty-four million pounds of relief supplies.[21] As the prime coordinator of this mission in Indonesia, *Abraham Lincoln* and its carrier strike group together "flew 1,800 sorties, delivered 2,700 tons of food, water and medicine and evacuated 3,000 people."[22]

While the sea bases were efficient, the logistics interface with the local government was not so smooth, making that aspect of the effort often chaotic and ad hoc. Tim Connolly, adviser to the World Food Program, reported at one point, "Planes are just dropping out of the sky unannounced."[23] Due to the confusion, there were concerns that aid might be sold instead of being given away free. Some Indonesians trusted the Americans more than their own soldiers to deliver the aid fairly. As Idris Rusli, who lived in a destroyed neighborhood of Banda Aceh, warned, "They [Indonesian personnel] will sell it themselves because they are very bad. . . . I worry about that. I really, really worry about that."[24] That points to a problem that may need to be addressed in the future: keeping U.S. military on sea bases leaves local government officials in control of the distribution of aid. As one senior agency official reported, "We've had some reports of TNI [Indonesia military personnel] . . . hoarding supplies—up to 30 per cent in some places."[25] Ambassador Pascoe later acknowledged, in response to a question from the press, that the "high level of corruption" in Indonesia was a "very serious problem."[26] Foreign coordination of humanitarian relief needs to take local corruption into account and take positive measures to deliver supplies directly to those who really need them the most.

Helicopter Access to Disaster Areas

Once humanitarian supplies were in-theater they had to be delivered, usually by helicopter. Andrew Natsios acknowledged, "I don't know where this relief operation would be without the ships, airplanes, helicopters, manpower, and can-do attitude provided by the U.S. military."[27] The emphasis on helicopters, however, put extraordinary demands on the pilots, their crew, and maintenance. At least eight pilots on *Abraham Lincoln* flew over a hundred hours, and maintenance personnel worked "unbelievable hours."[28]

Offloaded supplies awaiting distribution at Banda Aceh airport.

Even at that, had there been opposition on the ground the mission might not have succeeded as it did. As we have seen, simply dropping supplies from safe altitudes would not have been the answer. Brigadier General Paul J. Selva, operations director of the U.S. Transportation Command, Scott Air Force Base, later reflected that "air drops were considered, but food without water leads to dehydration. Water containers dropped from a plane tend to break."[29]

Americans fluent in the local languages were especially needed. People like Michael Bach, one of the USAID representatives, "made everything happen . . . spoke seven languages."[30] On many occasions a "language barrier between helos and the tower" arose, and the pilots would have to work by "sight rather than communication."[31] To interact with people on the ground, the helicopter crews would write up handbills with short messages, mainly safety related, such as STAY AWAY FROM HELICOPTERS, or DON'T THROW THINGS AT THE HELOS. They would then find a native speaker to make up a batch of these signs.[32]

Onload at Banda Aceh airport.

During UNIFIED ASSISTANCE, sea-base platforms were perhaps overly dependent on helicopter access. Had a hostile power wanted to disrupt the sea-base mission, the helicopters would have been an obvious target. Although such occasions were rare, helicopter pilots and their crews did sometimes feel threatened by desperate survivors. In future operations of this type, the sea-base ship should not become tied to helicopters solely but establish other forms of access as well.

Communications Promoting Efficient Coordination

Communications increased coordination but not necessarily command. Rear Admiral Crowder described his relationships with other militaries and organizations as "not a command relationship," since there was no "combined military chain of command." Each nation made its own agreement with Indonesia. The U.S. Navy helped by providing maps, pictures, and course rules for helicopters, but there was no "command relationship."[33] According to Admiral Fargo, it was "coordination vice command or compel"; still, he was convinced, "[you] must have some structure that allows you to prioritize."[34]

For many U.S. Navy officers, this type of coalition was completely new. In fact, as Captain Bob Aronson pointed out, "it was not a coalition at all." In hindsight, the mission coordinators should have set up an unclassified chat server, but even that would not have solved the problem of people not wanting to coordinate at all: "I wouldn't look at it like collaboration was a failure. . . . [I]t was a collection, not a coalition."[35]

Throughout UNIFIED ASSISTANCE, then, coordination with Indonesia remained key. The U.S. government communicated constantly with the Indonesian government through its embassy in Jakarta and the CSG-Indonesia headquarters in Medan. The absence of a status-of-forces agreement can severely constrain "the scope of the assistance provided by the foreign military." After the disaster, nations throughout Southeast Asia were urged to develop a regional network of status-of-forces agreements, based on a single standard document rather than a series of bilateral agreements.[36]

Finally, ship-to-shore contacts were often more difficult than global communications, and differences in networks—such as those between the unclassified NIPRNET and the classified SIPRNET—interfered with information sharing. The one exception was the information website made available on the Utapao Web server, which could be downloaded by anyone, virtually anywhere. This was one of the most information-friendly developments to come out of Utapao. What was required in addition was a similarly unclassified and open system that would have allowed personnel on ships or moving helicopters or in shore parties to communicate with each other.

Delivery of humanitarian supplies required coordination between Indonesians and Americans.

The Medical Mission

The comparatively late arrival of the USNS *Mercy* limited its potential benefits. *Mercy* did not treat patients until well over a month after the disaster, by which time most of the tsunami victims had either died or already received rudimentary medical attention. *Mercy* lacked the range of specialists available on other Navy ships and so needed constant support from them. For example, it had no helicopters of its own, and none of its crew were trained in how to operate a flight deck.

There were also some unexpected problems mixing civilians with naval personnel; in one case a Project Hope volunteer and a U.S. Navy sailor were discovered engaged in a sexual relationship. The sailor was returned to the parent command, while the Project Hope volunteer remained onboard. In another case, one of the Project Hope volunteers, who was openly gay, repeatedly complimented same-sex military personnel about their appearance.[37]

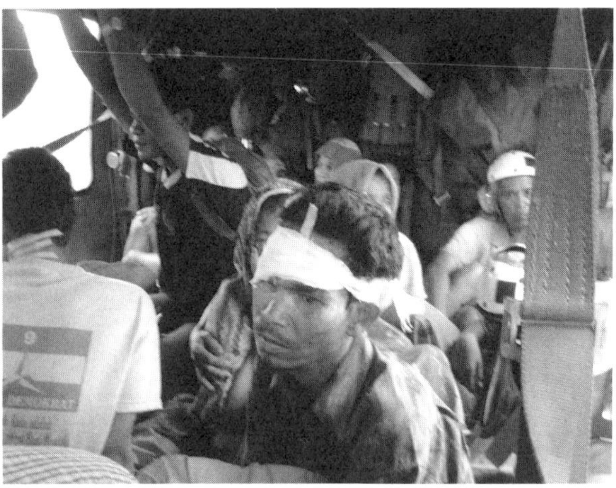

Helicopters delivered injured Indonesians directly to modern medical facilities.

Nevertheless, the arrival of *Mercy* and its extended civil-military medical mission to provide long-term health care sent a strong message to the Indonesian people and government. According to Captain Nathan Smith, *Mercy*'s mission was a successful combination of "Navy medicine, U.S. Public Heath Service, NGOs, and others." When the Indonesian military leader in Banda Aceh thanked *Mercy* there were "tears in his eyes," proving that a "hospital ship can be the best diplomat of the 21st century."[38]

Recommendations

In order to prepare for a future naval humanitarian mission similar to UNIFIED ASSISTANCE, several changes should be considered. As we have seen, insufficient and generally inadequate disaster reporting by the local governments and international media delayed accurate planning by many days, forcing commands to plan and execute simultaneously. New forms of information management are required as an adjunct to, perhaps even a substitute for, local reporting.

Sea-base platforms like *Abraham Lincoln* were criticized as only "marginally useful" compared to vessels able to function well in shallow, intertidal zones.[39] Although they provide the widest range of capabilities and represent high prestige, the size and operating cost of aircraft carriers would normally militate against their being sent as first responders to crises. In the speed-versus-capability equation, speed will almost always win out in a humanitarian disaster response. In future crises more mobile ships, including possibly the LCS, may prove faster and cheaper first responders.

Although the Transportation Command efficiently deployed prepositioned ships and moved many supplies by sea, the cost was high. In the future, air delivery in combination with air drops would be more cost-effective. The problem remains how to drop fresh water in bulk. While in the Indonesian case helicopter access appeared preferable, especially since local corruption might have otherwise undermined the proper delivery of supplies, such logistical problems will need to be examined and solved prior to future aid missions from the sea.

Helicopter pilots and crew were generally overworked and overstressed. Alternative delivery capabilities are clearly needed. Force protection was also a concern, since the helicopter pilots and their crews were generally unarmed and so vulnerable to attack.

Ship-based communications were particularly useful, especially between *Abraham Lincoln* and PACOM. However, regional and local communications were more difficult, and ship-to-shore communications could be problematic at best. Interestingly, communications problems seemed most acute not between various ships within a single strike group but among different strike groups trying to coordinate with each other.

Finally, while the medical mission embodied the best of intentions, it came on the scene far too late to help many of the sick and wounded. Faster response and greater mobility, perhaps through ultra-fast catamaran ships outfitted with containerized medical missions, should be seriously considered. Rapid supply of medicines specifically intended for a humanitarian mission would be easier if a modular system were adopted.

Despite various problems related to sea basing, logistics, access, communications, and medical assistance, the UNIFIED ASSISTANCE model worked very well in northern

Indonesia. Sea basing, by keeping American military forces away from civilian areas, especially at night, helped to decrease the chance of any cultural or religious friction, making this operational model particularly appropriate for future humanitarian missions in other highly sensitive regions. As it was, the positive impact on local public opinion was enormous. The operation had, in turn, a far-reaching impact on the global war on terror, increased good will between the American and Indonesian governments, fostered respect and appreciation by the Indonesian people, and so produced significant political benefits for the United States, both regionally and globally.

Notes

1. Epigraph from Richard C. Paddock, "Navy Airlifts Supplies to Ravaged Coast," *Los Angeles Times,* 3 January 2005.
2. Available at gopher.info.usaid.gov/press/speeches/2005/sp050118.html.
3. Slobodan Lekic, "Indonesia, Rebels OK Truce Talks: U.S. Calms Fear over Tsunami Aid," *Washington Times,* 24 January 2005.
4. Admiral Walter Doran (Ret.), interview with Elleman, Arlington, Virginia, 23 October 2006.
5. According to Admiral Thomas Fargo, many months prior to Katrina "I went back and briefed this thing, and we talked an awful lot of these things. Lessons spread quickly. . . . Value of getting these large-deck platforms into the area right off the bat." Admiral Thomas Fargo, telephone interview with Elleman, 30 July 2006.
6. Lieutenant Commander Jason Carter, CCSG-9 staff communications officer, Naval Historical Center [hereafter NHC] Oral History, 20 February 2005.
7. Fargo interview.
8. "Meeting the Enemy," *Baltimore Sun,* 10 January 2005.
9. "Interview with General Mike Hagee, Commandant, U.S. Marine Corps," *Defense News,* 7 February 2005.
10. Fargo interview.
11. CWO3 Bo Miller, CCSG-9 staff N312/helo element coordinator, NHC Oral History, 20 February 2005.
12. Commander Jon Ross, executive officer, VFA-151, NHC Oral History, 20 February 2005.
13. Rear Admiral Douglas Crowder, CCSG-9, NHC Oral History, 20 February 2005.
14. Mark S. Leavitt, Jeffrey M. Vorce, and Michael M. Hsu, "For Compassion and Country: UNIFIED ASSISTANCE," U.S. Naval Institute *Proceedings* 131, no. 4 (April 2005).
15. Ibid.
16. Craig Gima, "Pacific Commander Plans to Visit Devastated Region," *Honolulu Star-Bulletin,* 11 January 2005.
17. Fargo interview.
18. Carter oral history.
19. Ibid.
20. Erik Slavin, "General: Air Force Shifting Efforts toward Harder-Hit Areas in Indonesia," *Pacific Stars and Stripes,* 23 January 2005.
21. Available at usinfo.state.gov/gi/Archive/2005/Mar/05-433015.html.
22. William Choong, "U.S. Admiral Expects Better Bilateral Ties," *Singapore Straits Times,* 8 February 2005.
23. Paul Wiseman, "U.S. Troops Report Successes in Relief Effort: Governments around World Sending Help," *USA Today,* 12 January 2005.
24. Eric Unmacht, "View from U.S. Chopper," *Christian Science Monitor,* 6 January 2005.
25. David Ibison and Jake Lloyd-Smith, "Indonesia Delays U.S. Navy Aid Mission," *London Financial Times,* 10 January 2005.

26. Raymond Bonner, "U.S. Calls Indonesia Deadline for Troop Pullout Reasonable," *New York Times,* 14 January 2005.

27. Andrew S. Natsios, at U.S. Chamber of Commerce Forum on Post-Tsunami Reconstruction, 18 January 2005, available at gopher.info.usaid.gov/press/speeches/2005/sp050118.html.

28. HS-2, Group 1, NHC Oral History, 20 February 2005.

29. Robert Goodrich, "Relief Effort Dwarfs All for Scott Command, General Says," *St. Louis Post-Dispatch,* 7 January 2005.

30. Commander Chris Meyer, chief engineer, USS *Abraham Lincoln* (CVN 72), NHC Oral History, 20 February 2005.

31. HSL-47, Group 1, NHC Oral History, 20 February 2005.

32. Captain Bob Aronson, CCSG-9 staff N6/ACOS C41, NHC Oral History, 15 February 2005.

33. Crowder oral history.

34. Fargo interview.

35. Aronson oral history.

36. United Nations, *Report and Summary of Main Conclusions,* Regional Workshop on Lessons Learned and Best Practices in the Response to the Indian Ocean Tsunami, Medan, Indonesia, 13–14 June 2005 (New York: Office for the Coordination of Humanitarian Affairs, 5 July 2005), available at www.reliefnet.int.

37. Lieutenant Eric Molldrem, USNS *Mercy* (T-AH 19), NHC Oral History, 3 June 2005.

38. Captain Nathan Smith, USNS *Mercy* (T-AH 19), NHC Oral History, 6 June 2005.

39. Craig Hooper, "Building a Fleet for Coastal Rescue," *Boston Globe,* 4 January 2005.

CHAPTER NINE

The Political Benefits of UNIFIED ASSISTANCE

Some of our students who used to be quite aggressive [in their anti-American beliefs] have become more moderate now.

DR. FADIL LUBIS, STATE INSTITUTE OF ISLAMIC STUDIES
MEDAN, INDONESIA, 6 JANUARY 2005

Although the bulk of American troops were gone after only six weeks, UNIFIED ASSISTANCE dramatically improved U.S.-Indonesian government-to-government and military-to-military relations, and so furthered the goals of the global war on terror and of regional cooperation.[1] From the very beginning, the American response to the tsunami crisis set up conditions for improved relations. Secretary of Defense Donald Rumsfeld argued that renewing full U.S. military ties with the world's largest Muslim nation was an essential part of winning the war on terror, since traditionally Indonesians practice a moderate form of Islam.[2]

The final relief numbers provide compelling evidence of the impact of UNIFIED ASSISTANCE on Southeast Asia in general, and Indonesia specifically. The United States government donated $950 million; Australia was the second-largest aid provider, at $750 million in grants and loans, followed by Germany at about $680 million; Japan pledged $540 million, the European Union $624 million; China was far behind at $82 million, with Taiwan close on its heels at $50 million in aid. According to Andrew S. Natsios, this made the United States' pledge "the largest for a single disaster in the country's history." In addition, American citizens "have given $700 million to charities for the relief effort."[3]

But one senior Asian military official, who requested anonymity, warned the United States to be cautious in its attempts to improve relations with Indonesia too quickly. "I recognize from an American perspective, this is a great opportunity to restore their image and to make a real contribution. Nobody can match the resources the Americans are bringing in. But America does carry a certain baggage in these parts and it doesn't take much to lose goodwill, just a few obnoxious actions."[4] Sea basing, by keeping

American forces at arms length and out of sight, especially at night when visiting military personnel would normally have a chance to mingle with locals, allowed the United States to retain the Indonesian citizenry's good will, even while improving political ties with a key country in the fight against terrorism.[5]

AMCS Ray Adams and AD3 Jason Shireman with grateful Indonesian children, Kreung Sabe, Aceh Province.

Sea basing was also particularly appropriate in a country where Muslims predominate in the population. One newspaper pointed out that "with the exception of Indonesia, the hardest-hit areas are not Muslim. Sri Lanka's separatist Tamils are mostly Hindus and its ruling Sinhalese mostly Buddhists. India's southeast coast is largely Hindu, with many Christians. Most Thais are Buddhist, and in that country, about half of the tsunami's victims were foreigners. The wave virtually missed the heavily Muslim shorelines of Bangladesh and Malaysia."[6] U.S. troops could be based ashore without any apparent concern in Thailand and Sri Lanka, where the local cultures and religions were not antagonistic toward Americans, but in northern Indonesia sea basing was clearly a preferable option.

Once cultural tensions had been assuaged, the U.S. government could take political advantage of UNIFIED ASSISTANCE to halt its embargo of military goods to the Indonesian government and reopen normal diplomatic relations. One early example of improved military-military relations was the American decision to supply Indonesia with parts to repair their C-130s, assistance that had been denied to them previously, by the Clinton administration.[7] Prior to the arrival of these spare parts, only eight of Indonesia's twenty-five C-130s were fit to fly.[8]

Indonesian refugees waving goodbye to American aid providers.

During February 2005, Washington also resumed IMET, the International Military Education and Training program, for Indonesian military personnel, as well as the sale of nonlethal military equipment. Later, in May 2005, President Bush stated that "the resumption of normal military relations would be in the interest of both countries." Indonesian president Susilo Bambang Yudhoyono, who had graduated from the IMET program and once called the United States his "second home," supported this shift in American policy. On 22 November 2005, President Bush lifted an embargo on military exports and foreign military financing to Indonesia and allocated a million dollars in aid to the Indonesian Navy for 2006. This change in policy, argued Dr. Greg Fealy, an Australian National University lecturer, was primarily a product of UNIFIED ASSISTANCE, since the "tsunami spurred unprecedented cooperation with Indonesia's military."[9]

The creation of U.S.-Indonesian institutional bonds supporting the war on terror was a major by-product of the good working relations developed during the humanitarian relief operation. The sea-based humanitarian mission of OUA thus had a truly global effect. As Secretary of the Navy Donald C. Winter stated on 13 June 2006, "We have seen significantly positive impacts in Indonesia, Pakistan and the Horn of Africa as a direct result of our and other nations' humanitarian assistance and disaster relief."[10]

Within Asia, the presence of the U.S. military during UNIFIED ASSISTANCE also helped to reassure regional allies that a rapidly rising China would not slip in and fill a geopolitical vacuum. Prior to December 2004, the viability of the continued American presence in the region had been questioned, especially by the spectacular rise of China's "soft power."[11] The Chinese armed forces' logistics organization was called upon to help tsunami victims, and for the first time China played a role in assisting Indonesia, Sri Lanka, the Maldives, and Myanmar, but its total aid was low, especially in dollar terms—only $82 million in government aid and another $35 million in private donations. One author noted, "The government of Norway, by way of comparison, has pledged $180 million."[12]

Meanwhile, the U.S. decision to send "hard power" assets, like *Abraham Lincoln,* produced substantial "soft power" benefits.[13] Since UNIFIED ASSISTANCE, assessments of China's "soft power" throughout the region have subtly changed, with some scholars now arguing that "while China's soft power is increasing, Beijing faces serious constraints in translating these resources into desired foreign-policy outcomes."[14] Among other things, China's inability to support large numbers of humanitarian forces overseas undoubtedly reemphasized to the Southeast Asian governments the importance of retaining the U.S. military as the primary security provider.[15]

When viewed in terms of the Confucian concept of *ren,* or "humaneness," Washington was able to outshine Beijing by far. China is clearly aspiring to become the regional superpower by using a whole range of government powers, including its military forces, but when put to the test its naval forces failed. In recent articles discussing whether or not China should build its own aircraft carrier, Chinese researchers have highlighted a carrier's ability to conduct "disaster relief, and rebuilding."[16]

It was evident to all countries in the region that whereas the U.S. government could quickly dispatch a carrier strike group and an expeditionary strike group to assist in the humanitarian operations, and Japan tasked its maritime forces to assist, the Chinese did not order their navy to sea. China's humanitarian efforts following the tsunami were modest; according to Admiral Fargo, the Chinese "haven't quite figured out how to work effectively in these multinational or coalition situations. They don't have the structure or the training."[17] As Jeffrey Bader, a National Security Council Asian expert during the Clinton administration, noted, "This gives us an opportunity to remind the countries in the region that there are things that we can do that no one else can do—and, in particular, China can't do."[18]

This operation may have also sent an unintended military warning to China, since the USS *Abraham Lincoln*'s daily routine also included flight operations off the western mouth of the Malacca Straits. As Captain Raymond Ginetti later recounted, every night *Abraham Lincoln* would move about fifty miles out from shore, to return the next day and resume the humanitarian mission.[19] One reason *Abraham Lincoln* had to move so far offshore was that under U.S. Navy rules, carrier-based pilots who do not train (as pilots of the carrier's fixed-wing aircraft could not, under these conditions) for two weeks straight must undergo extensive retraining; the Indonesian government,

Commander Carrier Strike Group 9, Rear Admiral Douglas Crowder, points to sailors gathered on the flight deck while holding a framed photo of an Indonesian woman holding a sign that states "U.S. Soldiers Don't Leave."

however, was hesitant to allow the ship's fighter pilots to use Indonesian airspace. But if this training was conducted outside of Indonesia's sovereign waters, it took place in a highly strategic part of the world—the westernmost end of the Straits of Malacca—just where U.S. naval forces might choose to establish a blockade in a serious Sino-American conflict.[20] Therefore, the training missions clearly showed China how quickly U.S. forces could reach the area, and how efficiently they might be able to interdict Chinese trade. This lesson can be applied, in particular, to oil purchased from the Persian Gulf, 80 percent of which passes through the straits on its way to China.[21]

Prior to UNIFIED ASSISTANCE, the U.S. government's image in Indonesia clearly needed a boost. Following the invasion of Iraq, the popularity of the United States in Indonesia sank from 61 percent to only 15 within just a year, according to a poll by the Manhattan-based Pew Research Center. However, in one poll conducted after *Abraham Lincoln* and *Mercy* had left Indonesia the "results were a 39% favorable increase in how the Indonesian people view the United States. That's a significant increase, and this type of mission is wonderful for diplomacy."[22]

More importantly, in a survey taken in June 2005, 59 percent of the Indonesian people agreed that the United States paid a "great deal or a fair amount of attention to their country's interests," placing it among only four countries to hold that view (one of them being the United States itself) out of sixteen surveyed.[23] As R. William Liddle, an Indonesia scholar at Ohio State University, noted, during UNIFIED ASSISTANCE Internet postings in Indonesia became remarkably friendly, including one that read: "Welcome Uncle Sam!"[24]

Notes

1. Epigraph quoted in Barbara Slavin and Kathy Kiely, "Summit Today on $3 Billion in Aid," *USA Today*, 6 January 2005.
2. "*Lincoln* in Indonesia," *Wall Street Journal*, 4 January 2005.
3. Elizabeth Becker, "U.S. Nearly Triples Tsunami Aid Pledge, to $950 Million," *New York Times*, 10 February 2005.
4. Barbara Demick, "U.S. Troops Are Relieved to Be Greeted with Smiles," *Los Angeles Times*, 11 January 2005.
5. Neil King, Jr., and Donald Greenlees, "Aid Effort in Indonesia Could Lift U.S. Image in Eyes of Muslims," *Wall Street Journal*, 5 January 2005.
6. Donald G. McNeil, Jr., "Hearts, Minds and Flood Relief," *New York Times*, 9 January 2005.
7. Josh White, "Wolfowitz Cites Sri Lanka's Progress on Reconstruction: Regional Relief Efforts Called 'Platform' for Future," *Washington Post*, 18 January 2005.
8. Eric Schmitt, "U.S. and Indonesia May Restore Military Link," *New York Times*, 17 January 2005.
9. "Indonesia's Stature Rises with New Security Pact," *Christian Science Monitor*, 17 January 2006.
10. Richard Alexander, "SECNAV: We Must Think Like Our Enemies," *Newport Navalog*, 16 June 2006.
11. Eric Teo Chu Cheow, "Paying Tribute to Beijing; An Ancient Model for China's New Power," *International Herald Tribune*, 21 January 2004.
12. David Fullbrook, "China, the New Kid on the Charity Block," *International Herald Tribune*, 16 March 2005.
13. Vice Admiral John Morgan, "USN on the Challenges Facing the Navy/Nation," Intersessional Conference: Maritime Strategy Symposium 1, Naval War College, Newport, Rhode Island, 14 November 2006.
14. Bates Gill and Yanzhong Huang, "Sources and Limits of Chinese 'Soft Power,'" *Survival* 48, no. 2 (Summer 2006), pp. 17–36.
15. King and Greenlees, "Aid Effort in Indonesia Could Lift U.S. Image in Eyes of Muslims."
16. Andrew S. Erickson and Andrew R. Wilson, "China's Aircraft Carrier Dilemma," *Naval War College Review* 59, no. 4 (Autumn 2006), p. 29.
17. Admiral Thomas Fargo, telephone interview with Elleman, 30 July 2006.
18. Tom Raum, "U.S. May Boost Image if Efforts Seem Nonpolitical," *Philadelphia Inquirer*, 5 January 2005.
19. Captain Raymond Ginetti, navigator, USS *Abraham Lincoln* (CVN 72), Naval Historical Center Oral History, 20 February 2005.
20. Denis D. Gray, "Indonesia Denies U.S. Pilots Use of Airspace; Carrier Leaves," *Washington Times*, 13 January 2005.
21. Ioannis Gatsiounis, "Malacca Strait: Target for Terror," Asia Times Online, 11 August 2004, www.atimes.com/atimes/Southeast_Asia/FH11Ae02.html.
22. Letta Tayler, "When Relief Fights Belief," *Long Island Newsday*, 23 January 2005.
23. Pew Global Attitudes Project, 24 June 2005, available at pewglobal.org/reports/display.php?PageID=803.
24. King and Greenlees, "Aid Effort in Indonesia Could Lift U.S. Image in Eyes of Muslims."

Afterword: Maritime Strategy and Naval Responses to Nonmilitary Threats

> *Since men live upon land and not upon the sea, great issues between nations at war have always been decided—except in the rarest of cases—by what your army can do against your enemy's territory and national life, or else by the fear of what the fleet makes it possible for your army to do.*
>
> SIR JULIAN CORBETT, BRITISH NAVAL STRATEGIST, 1911

While Sir Julian Corbett accurately highlighted the traditional function of navies assigned to support land forces, it is perhaps not surprising that he did not foresee the need for navies to respond to nonmilitary emergencies, let alone large-scale humanitarian relief missions like UNIFIED ASSISTANCE.[1] Humanitarian and military operations have usually been considered mutually exclusive. Only at the end of World War II, and then only by default, were militaries used to engage in nation-building operations. The Berlin Airlift of 1948–49, for example, was a humanitarian relief mission in response to a direct military threat, as the Soviet Union blockaded access to Berlin. Operation UNIFIED ASSISTANCE, by contrast, responded to a purely natural disaster.

What assets or special capabilities can navies provide in times of war and peace that the other services cannot? Theorists of military strategy have focused on two naval missions: "the establishment of control of the sea" and "the exploitation of the control of the sea toward establishment of control on the land."[2] While Alfred Thayer Mahan emphasized the many dimensions of "control of the sea," Corbett, in his *Some Principles of Maritime Strategy,* provided a structurally complete analysis of sea power, one that the leaders of the U.S. Navy read with great interest and soon made part of its intellectual tradition.[3]

Historically, the navies of sea powers have been doctrinally most comfortable with sea-control missions that actually take place on and concern the sea itself. The mission of influencing events on the land from the sea has been more challenging. Arguably, the technology and methods needed to obtain this goal did not become truly available until the middle of the twentieth century, between the great European wars. The 6 June 1944 D-Day landings of vast armies, with their weapons, ammunition, and armor, were perhaps the ultimate proof that this broader kind of sea control could be obtained.

Yet many people at the time considered that development exceptional and unlikely to be repeated. In October 1949, General Omar N. Bradley, chairman of the Joint Chiefs of Staff, stated, "I predict that large-scale amphibious operations will never occur again," while President Harry S. Truman's secretary of defense, Louis A. Johnson, reportedly told Admiral Richard L. Conolly, "We'll never have any more amphibious operations. That does away with the Marine Corps. And the Air Force can do anything the Navy can do nowadays, so that does away with the Navy."[4]

The 15 September 1950 Inchon amphibious landing during the Korean War demonstrated again the importance to the land of sea control. UNIFIED ASSISTANCE'S sea basing further showed that supplies can be transported directly from the sea to temporary points of distribution ashore. As General Hagee later explained, with sea basing "you erase the line between the sea and the land." During World War II and in Korea, the American military first had "to build up a logistics on the beach before we could move inland." With sea basing, that was no longer necessary.[5]

Clearly, navies that enjoy both aspects of sea control now have leverage that other services do not have. Corbett's discussion of sea control focused on the importance of interrupting enemies' sea lines of communication (SLOCs) and denying them the ability to use the sea—in other words, the imposition of a policy of naval blockade against another country. He did not foresee that the reverse might some day become equally important: the *augmentation* of another country's maritime lines of communication, so as to deliver humanitarian supplies in times of need.

In this regard, Corbett's definition of a naval blockade embodied the twofold nature of sea control: "By occupying her [in this case the enemy's] maritime communications and closing the points of distribution in which they terminate we destroy the national life afloat."[6] This definition included three distinct goals: taking control over sea lines of communications, closing points of distribution, and destroying national life at sea and, by extension, on the land as well.

Naval humanitarian relief operations also appear to embody these three goals, but with the reverse impact in mind. In terms that are parallel but have precisely the opposite intended effects: By occupying the *affected country's* sea lines of communications and *keeping open* the points of distribution in which they terminate we *sustain* the national life. Many elements of naval humanitarian relief operations are therefore similar to those required for naval blockades, although naval relief missions hope to help whereas blockades aim to hurt. Instead of draining away logistical supplies, a sea base creates new and more robust logistical supply lines that reach temporary points of distribution in order to sustain a country's national life, even while restoring that country's infrastructure, trade, and commerce.

Of all military forces, only a navy can, by operating in the neutral world oceans, establish a "subtle, benign, ubiquitous presence" that gives it "a peacetime as well as a wartime employable usefulness to the nation."[7] From this perspective, naval relief missions can make use of many of the same "hard power" strategic assets as a naval blockade to produce significant "soft power" effects.[8]

* * * * * * *

What role have navies played to date in nonmilitary operations, particularly humanitarian relief missions? The primary functions of a navy can be "conceived as a trinity," embodying "the military, the diplomatic, and the policing functions." The first includes either the "threat" or the actual "use" of force. The second includes the use of naval forces in the "management of foreign policy short of the actual employment of force." The third, focusing on the internal matters of a state, is "rarely concerned with the armed forces of other states," but rather "with extending sovereignty over the state's own maritime frontiers."[9]

The one nonmilitary exception is nation building, in which naval forces can contribute to the internal stability and development of another country. Historically, the U.S. Navy has played major roles in humanitarian crises, beginning with its response to the 1906 earthquake in San Francisco. Soon afterward, in 1908, as noted above, the "Great White Fleet" assisted Messina, Sicily, after an earthquake and tsunami killed more than seventy thousand people. During World War II, the Lend-Lease program was another nonmilitary (at least, noncombat) naval operation that provided both humanitarian and military supplies.

Arguably, during the Cold War the humanitarian mission lapsed into secondary importance, the U.S. military being focused on the Soviet threat. Only after the collapse of the Soviet Union in 1991 was the Navy once again tasked to carry out high-priority humanitarian duties. Brigadier General Selva noted in 2005 that "the [U.S. Transportation] command has aided in 35 previous disasters worldwide since it was established in 1991."[10] Beginning in the spring and summer of 1991, the U.S. Navy received repeated humanitarian taskings. In May 1991, during Operation SEA ANGEL, American naval and Marine personnel assisted Bangladesh after Tropical Cyclone Marian killed between 139,000 and 152,000 people.[11] SEA ANGEL has been called a textbook civil-military operation in support of a humanitarian mission.[12]

During Operation FIERY VIGIL, in the wake of the Mount Pinatubo's 15 June 1991 eruption, a number of Seventh Fleet ships evacuated American military and families. The Navy's oldest aircraft carrier at the time, USS *Midway* (CV 41), and the newest, *Abraham Lincoln*, brought the evacuees to Cebu, from where they took Air Force and commercial planes to Andersen Air Force Base on Guam or to

Hickam in Hawaii, and then on to McChord Air Force Base in Washington or to Travis in California.

The joint U.S. Navy/Air Force Operation PROVIDE PROMISE (July 1992–March 1996) protected humanitarian relief efforts in the besieged cities of the former Yugoslavia. Admiral Jeremy M. Boorda, Commander in Chief, Allied Forces Europe, was responsible for supplying humanitarian relief to Bosnia-Herzegovina via air-land and air-drop missions. Beginning on 28 February 1993, the United States began an air drop of relief supplies aimed at Muslims surrounded by Serbian forces in Bosnia.

On 2 December 1996, President Bill Clinton announced the participation of U.S. military personnel and aircraft in Operation GUARDIAN ASSISTANCE to augment United Nations humanitarian assistance to refugees in Rwanda and the lakes region of eastern Zaire (in 1997, Zaire became the Democratic Republic of the Congo). This humanitarian relief mission involved U.S. naval aircraft overflying parts of the lakes region.

On 25 August 2000, President Clinton reported to Congress that the United States would participate in the UN Transitional Administration in East Timor. The U.S. military maintained personnel in East Timor to conduct humanitarian and civic assistance missions. In particular, Operation WARDEN included the use of four Marine Corps heavy-lift helicopters, an Army communications team, and 1,836 Marines and naval personnel afloat in the assault ship USS *Belleau Wood* (LHA 3).

In each of these humanitarian missions, the U.S. Navy participated in what would normally be called a "nation-building" operation. The Navy was not alone in these efforts, and indeed they were only possible through joint operations with the Army, Air Force, and Marine Corps. The naval contribution depended on the nature of the theater and the proximity to the sea. For geographical reasons, nation-building functions cannot be performed on an extensive scale by navies, but navies can make worthwhile contributions after natural disasters or during civil wars.

* * * * * * *

Based on an analysis of factors of time, space, and force, six naval humanitarian relief missions—to Bangladesh, the Philippines, Yugoslavia, Zaire, East Timor, and Indonesia—have certain unifying characteristics.

Naval humanitarian missions have many of the same characteristics as blockades. Like blockades they can be partial or total in terms of their overall ability to create temporary points of distribution to deliver and disperse supplies. Similarly, humanitarian missions can be distinguished as "near" or "far," in reference to the distance of the theater of operations from the country providing the naval platforms and the supplies, with UNIFIED ASSISTANCE being a good example of a "far" U.S. mission.

Due to the extensive earthquake and tsunami damage in Aceh Province, helicopters were essential for delivering humanitarian supplies from offshore sea bases to temporary distribution points far

The rate of implementation and duration of a humanitarian relief mission can influence effectiveness. Implementation can be rapid, gradually increasing, or intermittent, while the duration can be short, medium, or long. Rapidly deployed relief missions tend to be most suited to provide such basic necessities as water, food, and medicine, while gradual or intermittent relief missions are more suited for rebuilding basic services, such as roads, sanitation, communications, and long-term medical care.

The duration of a humanitarian mission can also indicate what its main focus will be. In UNIFIED ASSISTANCE, which lasted only about six weeks, the primary goal was to provide life-sustaining water, food, and medicines. Due to severe time constraints, U.S. military forces did not emphasize rebuilding basic infrastructure, although the hospital ship *Mercy* was able to assist in rebuilding Indonesia's emergency medical services.

Geographical limitations include access to the sea, riverine access inland, and the transportation infrastructure on land. Aid relief via naval vessels makes economic sense only under certain circumstances. For the U.S. Navy to take the lead role in a humanitarian relief operation, the afflicted country must be relatively isolated from its neighbors, especially neighbors that share contiguous land borders near the site of the disaster. Otherwise, relief is more efficiently delivered overland. For obvious reasons, island, peninsular, or coastal countries and those that are separated from their neighbors by natural obstacles, such as high mountain chains or extensive bodies of water, including lakes and rivers, are the most susceptible to naval assistance. The nation must also have ample landing areas for helicopter access—which was not an issue in Indonesia, due to the extreme damage.

Finally, countries with minimal transportation infrastructures, or more highly developed countries that have suffered catastrophic devastation, potentially benefit the most from the basic services that a modern naval force is able to deliver. In cases of internal natural disasters (such as earthquakes) where crucial transportation and communication infrastructure is destroyed, or large-scale coastal disasters (including tsunamis and typhoons) where boats, docks, and dock equipment are destroyed, responders must provide their own logistical flow, access, communications, and medical services, and only sea-based naval forces can do so.

After time and space, force constitutes perhaps the most critical dimension of humanitarian relief efforts. In the modern era, the ability to conduct a naval humanitarian mission requires surface ships, fixed-wing airplanes, and helicopters; joint or combined operations among naval, land, and, increasingly, air assets; and delivery of humanitarian supplies directly into the affected country's sovereign land, sea, or air space. The duration of the humanitarian mission also has an impact on force levels, since protraction can either deplete a fixed force or require reinforcements.

In the past, humanitarian missions conducted from the sea in the absence of major land operations were rare, usually involving simple air drops. Many recent naval humanitarian missions, like the Mount Pinatubo evacuation and OUA, have entailed joint and combined operations coordinating sea, land, and air operations. As fixed-wing aircraft and helicopters have become more available and dependable, they have been used in place of naval or land forces to deliver needed supplies.

Because fresh water had to be delivered along with food, air drops from planes were not an option in Aceh Province, requiring aid delivery by helicopter.

Technological breakthroughs have greatly influenced the cost, execution, and feasibility of all types of humanitarian relief missions from the sea. Joint and combined operations have also played an increasingly important role, with sea-air operations substituting for land-sea operations in the modern period. This has limited the size of the footprint ashore. Humanitarian relief missions by sea can also be executed unilaterally or in combination with a coalition. UNIFIED ASSISTANCE showed that some missions might also work best with a "collection" rather than a formal coalition, although without proper communications such an assemblage can rapidly devolve into chaos.

Humanitarian missions are a means to an end. At the operational level, they are simply a way to assist with transportation and communications so as to enable the flow of aid. At the strategic level, however, they can—sometimes alone but more often in combination with other actions—provide a means to achieve an overarching national goal that provides the rationale for providing assistance. Strategic goals can include, but are not limited to, improving diplomatic relations, increasing military-to-military cooperation, and fostering public good will. As this study has sought to show, UNIFIED ASSISTANCE clearly attained all of these strategic goals in the relations of the United States with Indonesia.

* * * * * * *

Evaluating the effectiveness of a humanitarian relief mission has three parts: Did the mission achieve its operational goal? Did this contribute to the achievement of strategic success? Were the benefits commensurate to the costs? Factors influencing effectiveness include availability of alternative lines of communication and of cheaper substitutes for relief supplies, and the size of the area receiving humanitarian assistance.

Table 2 shows that historically the most effective naval humanitarian relief missions focused on isolated regions that could readily be accessed only by the sea, were implemented rapidly, and were of short to medium duration. Of these six humanitarian relief efforts, three responded to civil wars—Yugoslavia, Zaire, and East Timor—while the other three were the result of natural disasters—Bangladesh, Pinatubo, and Indonesia. Success rates for the three natural disasters appear to be higher overall than for those undergoing civil wars, with East Timor being one possible exception, perhaps because of its relatively small geographic size and the operation's medium duration.

The absence of organized security threats to humanitarian forces following natural disasters helps make them succeed, while natural disasters tend to make afflicted populations extraordinarily receptive to donor nations. However, unlike the two other cases of natural disaster, Bangladesh and the Philippines, the Indonesian case provides a model for humanitarian aid in a disaster area that also has an active insurgency.

TABLE 2
U.S. Naval Relief Operations, 1991–2005

HUMANITARIAN MISSION	ALT. LAND ROUTES	GEOGRAPHIC COVERAGE	RATE OF IMPLEMENTATION	DURATION	OPERATIONAL/ STRATEGIC EFFECT
Bangladesh	none	huge	rapid	short	operational—high
Philippines	none	huge	rapid	short	operational—high
Yugoslavia	land lines	huge	intermittent	long	both—medium
Zaire	land lines	huge	rapid	medium	operational—medium
East Timor	none	limited	rapid	medium	both—medium
Indonesia	none	huge	rapid	short	both—high

UNIFIED ASSISTANCE was arguably the first American naval humanitarian mission conducted successfully despite a constant need for force protection.

Another important inhibiting factor in Yugoslavia and Zaire was the existence of alternative land lines, for which reason naval relief forces could play only a relatively small role, even over the long and medium terms. Reasons for lower success rates also appear to be connected to the size of the area involved, which in turn is often connected to the nature of the theater. In five of the six cases—East Timor being the only exception—the affected areas were huge.

The effectiveness of sea-based humanitarian relief missions is a function of the size of the affected area, the rate of implementation, and the level of cooperation of the local populations. The smaller the region and the less interconnected it is by land, the greater the impact of a naval power. Thus, the most obviously successful naval humanitarian relief missions involved sea powers assisting islands or isolated areas, such as Aceh Province, where the operation had the dual effect of replacing basic infrastructure and communications even while assisting people in need. The four most successful naval operations—Bangladesh, Pinatubo, East Timor, and Indonesia—closely fit these characteristics. Rear Admiral Crowder has effectively acknowledged this point: "This is a once in a generation, if not once in a lifetime, natural disaster that has occurred in this area of the world."[13]

As technology has changed, so too have the ways in which such missions acquire access to the afflicted countries. Over time, naval relief missions have shifted from sending surface ships to access port facilities to relying on aviation. Thus, helicopters played a particularly important role in UNIFIED ASSISTANCE. With the coming of the V-22 tilt-rotor, a new era of airborne relief missions may be imminent.

Future relief efforts from the sea will probably be conducted by multinational coalitions: the projected "thousand-ship navy," informal "coalitions of the willing," or UN-sponsored groupings. In such a multinational atmosphere, communications, including unclassified cell phones, Internet chat rooms, and satellite communications, will play ever more important roles. Given these changing circumstances, seapower will remain essential for conducting humanitarian aid missions, even though access and communications platforms will more often be in the air or even in space, as the importance of helicopters and satellite communications during OUA showed.

* * * * * * *

During UNIFIED ASSISTANCE, aircraft carriers, with numerous helicopters, and helicopter-capable amphibious ships were particularly effective. Although not a particularly cheap means for delivering aid, these "floating cities" provided many of the basic supplies required in the initial stage of the humanitarian relief operation, including fresh water, food, medical supplies, and building materials for temporary shelters, as well as in the widest variety of humanitarian activities on land.

To support the carriers and amphibious ships, forward logistical bases like the landing field at Utapao and the U.S. Navy's logistics center at Singapore worked closely with naval logistics ships provided by the Military Sealift Command. Together they transported enormous quantities of relief supplies, as well as the manpower and equipment needed to offload them. MSC ships provided an invaluable service, especially in the initial phases of the operation before the opening of reliable air distribution routes.

Due to the crippling damage to the infrastructure of the affected countries, as well as to the general lack of functioning central government organs and to widespread corruption, it was crucial that fixed-wing and rotary-wing platforms take relief supplies directly to the affected areas, avoiding the more normal permanent points of distribution in favor of temporary points.

Onloading tents from the United Nations High Commissioner for Refugees at Banda Aceh airport.

Communications were an essential part of the mission, both up and down a vertical chain of command linking the sea-base ships with the CSF headquarters in Utapao and PACOM in Honolulu, Hawaii, but also horizontally, linking all of the ships participating in the operation, including a wide range of non-U.S. naval and air assets, NGOs, and other contributing nations. Continuous contact between the sea bases and the aerial platforms was also a basic requirement, as was communication with the parties on shore.

Naval hospital ships, like the USNS *Mercy,* represented an extremely important second-wave capability by providing world-class medical care in case of epidemics. Not only were these vessels self-sustaining when it came to water, light, and electricity, but they were well stocked with medical supplies. Helicopters were able to fly patients directly from shore to the ship for tests, operations, or long-term medical care.

The UNIFIED ASSISTANCE model appears to be particularly useful in situations where the normal functions of governments and local service providers have been disrupted or when local conditions—due to cultural and religious differences—make sea-basing American military personnel preferable. Otherwise, these extremely costly sea-based capabilities can be redundant. The price disparity between local goods and those provided by the United States suggests that while navies might be one of the best first responders in a wide variety of humanitarian disasters, they will be most effective only when the disaster in question creates large-scale, systemic failures of central and local government services.

The strength of sea basing is in combining mobility, reliability, and flexibility at sea with cultural and political sensitivity on land. In OUA, mobility was necessary to bring the naval platforms to the scene of the disaster. However, in contrast to the time it took for the CSF 536 headquarters to stand up at Utapao, reliability (in terms of logistics and the access provided by fixed-wing and helicopter operations) and even more importantly flexibility (of command, control, and communications networks) allowed *Abraham Lincoln* to begin humanitarian operations quickly once it reached Indonesian waters.

Key capabilities in naval humanitarian relief missions include aircraft carriers or amphibious ships providing sea bases that can be positioned to reach deeply into afflicted regions; naval logistics ships and cargo airplanes that can transport enormous quantities of manpower, equipment, and humanitarian supplies uninterruptedly over long SLOCs; helicopters and fixed-wing airplanes that can fly in almost any weather and land where needed, not just on airfields; communications networks capable of linking a wide range of American and other naval ships, with shore stations, and the local authorities; and naval hospital ships that are linked by helicopter lift and can provide world-class medical care.

During the nineteenth and most of the twentieth centuries, the very thought that sea powers might regularly use naval platforms to deliver humanitarian aid, as opposed to cutting off and starving an enemy's supply lines, would have seemed alien. In the twenty-first century, however, national power and prestige are more and more characterized by "soft power." UNIFIED ASSISTANCE showed that "hard power" assets like aircraft carriers can also be the best providers of "soft power." This lesson is especially relevant as the "thousand-ship navy" is taking shape. The UNIFIED ASSISTANCE model developed in northern Indonesia was a real-life example of how such a "soft power" humanitarian relief operation can work in practice.

Notes

1. Epigraph in Sir Julian S. Corbett, *Some Principles of Maritime Strategy* (London: Conway Maritime, 1972), p. 14.
2. J. C. Wylie, *Military Strategy: A General Theory of Power Control* (New Brunswick, N.J.: Rutgers Univ. Press, 1989, repr. 1967), p. 39.
3. John B. Hattendorf, "Mahan Is Not Enough: Conference Themes and Issues," in *Mahan Is Not Enough: The Proceedings of a Conference on the Works of Sir Julian Corbett and Admiral Sir Herbert Richmond*, ed. James Goldrick and John B. Hattendorf (Newport, R.I.: Naval War College Press, 1993), p. 9.
4. Victor H. Krulak, "You Can't Get There from Here: The Inchon Story," available at www.usna.com/News_pubs/Publications/Shipmate/2000/2000_06/inchon.htm.
5. "Interview with General Mike Hagee, Commandant, U.S. Marine Corps," *Defense News*, 7 February 2005.
6. See Corbett, *Some Principles of Maritime Strategy*, as reprinted at gwydir.demon.co.uk/PG/Corbett/corbett.htm#Page_91, 95.
7. J. C. Wylie, "Mahan: Then and Now," in *The Influence of History on Mahan*, ed. John B. Hattendorf (Newport, R.I.: Naval War College Press, 1991), pp. 41–42.
8. Bruce Elleman and S.C.M. Paine, eds., *Naval Blockade and Seapower: Strategies and Counter-strategies, 1805–2005* (London: Routledge, 2006).
9. Ken Booth, *Navies and Foreign Policy* (New York: Holmes and Meier, 1979), pp. 15–17.
10. Robert Goodrich, "Relief Effort Dwarfs All for Scott Command, General Says," *St. Louis Post-Dispatch*, 7 January 2005.
11. Gary W. Anderson, *Operation Sea Angel: A Retrospective on the 1991 Humanitarian Relief Operation in Bangladesh* (Newport, R.I.: U.S. Naval War College, Center for Naval Warfare Studies, Strategy and Campaign Department, 1992).
12. Chris Seiple, *The U.S. Military/NGO Relationship in Humanitarian Interventions* (Carlisle Barracks, Pa.: U.S. Army War College, Peacekeeping Institute, Center for Strategic Leadership, 1996).
13. Alan Sipress and Noor Huda Ismail, "Relief Transcends U.S.-Indonesia Divide: Rights Concerns Sidestepped as Militaries Respond Jointly to Disaster," *Washington Post*, 4 January 2005.

Appendix: Chronology of the Tsunami Disaster and Humanitarian Response

All dates are U.S. time.

2004	**26 December:** 9.0 earthquake/tsunami
	28 December: C7F directs CCSG-9 to get under way from Hong Kong
	29 December: JTF-536 HQ established at Utapao, Thailand
	30 December: MPSRon ordered to sortie from Guam
2005	**1 January:** CSG-9 arrives Banda Aceh; USS *Abraham Lincoln* provides aid
	2 January: CJTF-536 arrives Utapao
	January: Combined Support Group Indonesia (CSG-Indonesia) HQ stood up in Medan, Sumatra ESG-5 arrives Medan; USS *Bonhomme Richard* takes on supplies
	6/7 January: CSF-536 HQ establishes SIPRNET/NIPRNET connectivity NavFor exceeds 500,000 pounds aid delivered (250 tons) ESG-5 repositions IVO Meulobah, begins delivering aid
	8 January: USNS *Mercy* under way from San Diego
	9 January: Thailand and Sri Lanka move to recovery phase; Indonesia remains in crisis phase
	10 January: ESG-5 begins LCAC ops IVO Meulobah NavFor passes one million pounds aid delivered (five hundred tons)
	14 January: First UN assessment teams go ashore from *Abraham Lincoln* Naval forces pass two million pounds aid delivered (one thousand tons)
	15 January: CSG-Indonesia in Medan up on SIPRNET
	16 January: CSG-Indonesia up on VTC
	18 January: CSG-Indonesia FWD in Banda Aceh up on SIPRNET and NIPRNET ESG-5 departs, USS *Essex* arrives CCSG-9 assumes duties as CSG-Indonesia Naval forces pass three million pounds aid delivered (1,500 tons)
	20 January: First coordination meeting among CSG-Indonesia, UN, USAID, NGOs CSG-Indonesia creates SPARK Team to facilitate NGOs SA
	22 January: Combined Support Group–Thailand disestablished Naval forces pass four million pounds aid delivered (two thousand tons)
	25 January: All U.S. forces out of Medan
	28 January: All U.S. forces out of Sabang Naval forces pass five million pounds aid delivered (2,500 tons)
	29 January: Combined Support Group–Sri Lanka disestablished

2005	**30 January:** USNS *Mercy* arrives Singapore
	3 February: *Mercy* relieves *Abraham Lincoln* CSG, which returns to United States
	9 February: *Essex* completes relief ops, proceeds to Persian Gulf
	12 February: CSF-536 in Utapao is disestablished
	14 February: Naval forces pass 9.5 million pounds of aid delivered (4,750 tons)
	16 March: *Mercy* departs Indonesia, en route to Singapore

List of Abbreviations and Acronyms

A

ABFC chief aviation boatswain's mate (fuels)

ACB amphibious construction battalion

ACOS assistant chief of staff

ADF Australian Defence Force

AFMIC Armed Forces Medical Intelligence Center

AIP Antisurface Warfare Improvement Program

AIS Automatic Identification System

AMC Air Mobility Command

AO1 Aviation Ordnanceman 1st Class

APAN Asia Pacific Area Network

APS afloat prepositioning shipping

B

B2C "Bravo to Sea" [embarkation of an additional helicopter squadron in a CSG]

C

C2 command and control

C3 command, control, and communications

C4I command, control, communications, computers, and intelligence

C7F Commander, Seventh Fleet

CAG civil affairs group; carrier air group commander

CAT crisis action team

CCC Combined Coordination Center [Utapao, Thailand]

CCSG Commander, Carrier Strike Group

CDRC Combined Disaster Relief Center [used instead of CMOC]

CENTRIXS Collaborative Enterprise Regional Information Exchange System

CFAC	combined force air component
CFACC	combined force air component commander
CG	commanding general
CJTF	combined joint task force
CLF	Combat Logistics Force
CMOC	Civil Military Operations Center
CNA	Center for Naval Analyses
CNO	Chief of Naval Operations
CODS	carrier onboard delivery [aircraft]
COS	chief of staff
CPR	cardiopulmonary resuscitation
CSF	combined support force [in UNIFIED ASSISTANCE, the JTF became a CSF]
CSG	carrier strike group; combined support group
CSG-9	Carrier Strike Group 9
CSL	cooperative security location
CSO	chief staff officer
CTF	Commander, Task Force
CVW	carrier air wing
CWO3	Chief Warrant Officer 3

D

DART	disaster assistance response team [USAID]
DCAG	deputy carrier air group commander
DesRon	destroyer squadron
DET	detachment
DHS	demographic and health survey
DIA	Defense Intelligence Agency
DoD	U.S. Department of Defense

DR	disaster relief
DRAT	disaster relief assessment team [U.S. Navy]
DRRS	Defense Readiness Reporting System

E

EEI	essential elements of information
EOD	explosive ordnance disposal
EODMU	explosive ordnance demolition mobile unit
ESG	expeditionary strike group
ESG-5	Expeditionary Strike Group 5

F

FAST	fleet antiterrorism strike team
FCE	Forward Command Element
FEC	facilities engineering center
FHA	foreign humanitarian assistance
FISC	fleet and industrial supply center
FLOT	forward line of own troops
FSSG	force service support group

G

GAM	Gerakan Aceh Merdeka [Free Aceh Movement]

H

HA	humanitarian assistance
HA/DR	humanitarian assistance/disaster relief
HM2	Hospital Corpsman 2nd Class
HOC	humanitarian operations center
HQ	headquarters
HS	helicopter antisubmarine squadron
HSL	helicopter antisubmarine squadron light [U.S. Navy]
HSS	health service support
HSV	high-speed vessel

I

IMET	International Military Education and Training
IOM	International Organization of Migration
IR	intelligence requirement
ISR	intelligence, surveillance, and reconnaissance
IT1	Information Technology Technician 1st Class
ITC	Chief Information Technology Technician

J

J/NMETL	joint/Navy [or naval] mission-essential task list
JCS	Joint Chiefs of Staff
JFAC	joint force air component
JFMCC	joint force maritime component commander
JICPAC	Joint Intelligence Center Pacific
JMETL	joint mission-essential task list
JMFU	joint meteorological and oceanographic forecast unit
JPAC	Joint Prisoners of War/Missing in Action Accounting Command
JTF	joint task force

L

LCAC	landing craft, air cushion
LCS	Littoral Combat Ships
LCU	landing craft, utility
LFA	lead federal agency
LNO	liaison officer
LOCC	logistics operations coordination center
LTTE	Liberation Tigers of Tamil Eelem [Tamil Tigers]

M

MA1	Master-at-Arms 1st Class
MarDiv	Marine division
medevac	medical evacuation
MEF	Marine expeditionary force

METL	mission-essential task list
METOC	meteorological and oceanographic
MEU	Marine expeditionary unit
MNF	multinational force
MPA	maritime patrol air
MPAT	Multinational Planning Augmentation Team
MPF	maritime prepositioning force
MPSRon	maritime prepositioning ship squadron
MRI	magnetic resonance imaging
MSC	Military Sealift Command
MSD	mobile security detachment
MSS	mobile security squadron
MTF	medical treatment facility
MUSE	mobile utility support equipment

N

NavFac	naval facility
NEHC	Navy Environmental Health Center
NEPMU	Navy environmental and preventive medicine unit
NGO	nongovernmental organization
NIPRNET	Non-Secure Internet Protocol Router Network
NLL	Navy lesson learned
NLLS	Navy Lessons Learned System
NMETL	Navy [or naval] mission-essential task list
NTA	Navy [or naval] tactical task
NTIMS	Navy Training Information Management System
NTTL	Navy [or naval] Tactical Task List
NWDC	Navy Warfare Development Command

O

OCHA	Office for the Coordination of Humanitarian Affairs
OFDA	Office of U.S. Foreign Disaster Assistance
OPDS	Offshore Petroleum Discharge System
optempo	operating tempo
OSD	Office of the Secretary of Defense
OUA	Operation UNIFIED ASSISTANCE

P

PACOM	Pacific Command
PAO	public affairs officer
POM	preparation for overseas movement
PTWC	Pacific Tsunami Warning Center
PWPT	potable-water pillow tank

R

RFID	radio frequency identification
RMSI	Regional Maritime Security Initiative
ROA	relief operations area
ROWPU	reverse-osmosis water purification unit
RSAF	Republic of Singapore Armed Forces

S

SAF	Sri Lankan Armed Forces
SIPRNET	Secret Internet Protocol Router Network
SJA	staff judge advocate
SJFHQ	Standing Joint Force Headquarters
SLOC	sea line of communication
SOP	standard operating procedure
STAR	scheduled theater airlift route

T

T-AH	hospital ship [MSC]
TACMEMO	tactical memorandum
TRAP	tactical recovery of aircraft and personnel

TSC	theater security cooperation
U	
UAV	unmanned aerial vehicle
UCT	underwater construction team
UJTL	Universal Joint Task List
UN	United Nations
UNICEF	United Nations Children's Fund
UNJLC	United Nations Joint Logistics Center
USAID	U.S. Agency for International Development
USAID/OFDA	U.S. Agency for International Development/Office of Foreign Disaster Assistance
USDAO	U.S. Defense Attaché Office
USG	U.S. government
USMC	U.S. Marine Corps
USN	U.S. Navy
USSG	U.S. Support Group
UTC	Coordinated Universal Time [effectively, Greenwich Mean Time]
V	
VAQ	tactical electronic warfare squadron
VFA	fighter squadron
VTC	video teleconference
W	
WFP	World Food Program [UN]
WHO	World Health Organization
Y	
YN2	Yeoman 2nd Class

Bibliography

NAVAL HISTORICAL CENTER, ORAL INTERVIEW SUMMARY, NCDD 206

Aronson, Capt. Bob, CCSG-9 staff N6/ACOS C41, 15 February 2005.

Baca, Cdr. John, gunnery officer and Weapons Department head, USS *Abraham Lincoln* (CVN 72), 20 February 2005.

Banerjee, Sunoy, PhD, CNA representative, USS *Abraham Lincoln* (CVN 72), 20 February 2005.

Bemiller, Capt. Timothy, USNS *Mercy* (T-AH 19), 7 June 2005.

Boyles, Cdr. Dan, executive officer, HSL-47, 20 February 2005.

Campbell, Capt. Kevin, CCSG-9 staff N5, 20 February 2005.

Campbell, Cdr. Marquez, USNS *Mercy* (T-AH 19), 7 June 2005.

Carter, Lt. Cdr. Jason, CCSG-9 staff communications officer, 20 February 2005.

Clark, Lt. Cdr. Suzanne, USNS *Mercy* (T-AH 19), 6 June 2005.

Clevenger, Lt. Sean, CAT Cell member, CSG-9, 22 February 2005.

Comlish, Cdr. Jean, USNS *Mercy* (T-AH 19), 3 June 2005.

Corpuz, CWO2 Leo, USNS *Mercy* (T-AH 19), 4 June 2005.

Crowder, Rear Adm. Doug, CCSG-9, 20 February 2005.

Daniels, Lt. Cdr. John, PAO, USS *Abraham Lincoln* (CVN 72), 20 February 2005.

Ferrera, Lt. Cdr. Stephen, head of radiology, USNS *Mercy* (T-AH 19), 5 June 2005.

Gangnon, QM1 Roderic, EODMU, 11 Group, 20 February 2005.

Garcia, Cdr. Jamie, USNS *Mercy* (T-AH 19), 4 June 2005.

Ginetti, Capt. Raymond, navigator, USS *Abraham Lincoln* (CVN 72), 20 February 2005.

Horan, Cdr. Michael, commanding officer, HS-2, 20 February 2005.

HS-2, Group 1, 20 February 2005.

HS-2, Group 2, 20 February 2005.

HSL-47, Group 1, 20 February 2005.

Hsu, Lt. Cdr. Michael, LNO, USS *Abraham Lincoln* (CVN 72), 20 February 2005.

Hughes, Cdr. Ron, strike operations officer, USS *Abraham Lincoln* (CVN 72), 20 February 2005.

Hummeldorf, Cdr. Kurt, USNS *Mercy* (T-AH 19), 3 June 2005.

Kaufmann, Capt. Jon, Commander, DesRon 9, 20 February 2005.

Kiser, Capt. Bob (Fleet Surgeon), Capt. Rick Morrison, Lt. Cdr. Paul Miller, Lt. Cdr. Paul Miller, Lt. Cdr. Steve Jeffs, ComPACFLT., Pearl Harbor, Hawaii, 25 February 2005.

Klunder, Capt. Matthew, DCAG, CVW-2, 20 February 2005.

Konzek, YN2 Daniel, USNS *Mercy* (T-AH 19), 4 June 2005.

Llewellyn, Capt. David, commanding officer, USNS *Mercy* (T-AH 19), 2 June 2005.

McCulley, Capt. Tim, MSC Liaison, PACFLT Staff, Pearl Harbor, Hawaii, NHC, 2 March 2005.

McCully, Capt. Tim, USNS *Mercy* (T-AH 19), 7 June 2005

McDonald, Cdr. Karen, USNS *Mercy* (T-AH 19), 6 June 2005.

McManus, Lt. Christine, DesRon 9 staff air operations officer, 15 February 2005.

Medical Group 1, USS *Abraham Lincoln* (CVN 72), 20 February 2005.

Meyer, Cdr. Chris, chief engineer, USS *Abraham Lincoln* (CVN 72), 20 February 2005.

Michael, Cdr. Frank, commanding officer, HSL-47, 20 February 2005.

Miller, CWO3 Bo, CCSG-9 staff N312/helo element coordinator, 20 February 2005.

Molldrem, Lt. Eric, USNS *Mercy* (T-AH 19), 3 June 2005.

Moore, HM2 Jason, USNS *Mercy* (T-AH 19), 5 June 2005.

Moss, Lt. Cdr. Joel, HS-2, Group 1, 20 February 2005.

Murdock, Lt. Cdr. Scott, DesRon 9 chief staff officer, 20 February 2005.

Ohl, Lt. Cdr. Tony, CCSG-9 staff N5A, 20 February 2005.

Owen, Lt. Cdr. John, USNS *Mercy* (T-AH 19), 4 June 2005.

Parker, Lt. Cdr. John, USNS *Mercy* (T-AH 19), 5 June 2005.

Parra, AO1 Kevin, EODMU, 11 Group, 20 February 2005.

Quidachay, Cdr. Vince, CCSG-9 staff N4, 20 February 2005.

Roby, Capt. Brian, CCSG-9 chief of staff, 20 February 2005.

Ross, Cdr. Jon, executive officer, VFA-151, 20 February 2005.

Salata, Lt. Cdr. Jason, PAO, PaCom, Pearl Harbor, Hawaii, 2 March 2005.

Senko, Lt. Robert, USNS *Mercy* (T-AH 19), 7 June 2005.

Serbas, Lt. Cdr. Jeosalina, USNS *Mercy* (T-AH 19), 5 June 2005.

Smith, Capt. Nathan, USNS *Mercy* (T-AH 19), 6 June 2005.

Solar, ABFC Windsor, USNS *Mercy* (T-AH 19), 5 June 2005.

Supply Department, CCSG-9, 20 February 2005.

Thompson, Cdr. Russ, executive officer, HS-2, 20 February 2005.

Threet, Lt. Cdr. Erik, ITC Bobbi Jo Williams, IT1 Joshua Davis, IT1 John Opsteegh, and IT1 Damion Jasso, USNS *Mercy* (T-AH 19), 6 June 2005.

Valle, MA1 Scot, USNS *Mercy* (T-AH 19), 6 June 2005.

Verheul, Cdr. Dan, CCSG-9 staff N2/ACOS Intelligence, 20 February 2005.

Walcott, Lt. Cdr. Tom, chaplain, CVW-2, 20 February 2005.

Williams, Cdr. Ted, executive officer, VAQ-131, 20 February 2005.

OTHER PRIMARY SOURCES

Military Sealift Command (MSC), Tsunami Relief Effort records.

SECONDARY SOURCES

Anderson, Gary W. *Operation Sea Angel: A Retrospective on the 1991 Humanitarian Relief Operation in Bangladesh.* Newport, R.I.: U.S. Naval War College, Center for Naval Warfare Studies, Strategy and Campaign Department, 1992.

Booth, Ken. *Navies and Foreign Policy.* New York: Holmes and Meier, 1979.

Corbett, Sir Julian S. *Some Principles of Maritime Strategy.* London: Conway Maritime, 1972.

Hattendorf, John B. "Mahan Is Not Enough: Conference Themes and Issues." In *Mahan Is Not Enough: The Proceedings of a Conference on the Works of Sir Julian Corbett and Admiral Sir Herbert Richmond,* edited by James Goldrick and John B. Hattendorf. Newport, R.I.: Naval War College Press, 1993.

Luke, Ivan. "Operation UNIFIED ASSISTANCE Case Study." NWC 3096, A-3. Newport, R.I.: Naval War College, n.d.

Military Sealift Command. "Operation UNIFIED ASSISTANCE (OUA): Southeast Asia Earthquake and Tsunami Relief Effort (2004–2005)."

"National Plan to Achieve Maritime Domain Awareness: For the National Strategy For Maritime Security." Washington, D.C.: October 2004.

Navy Warfare Development Command. "Humanitarian Assistance/Disaster Relief (HA/DR) Operations Planning." TACMEMO 3-07.6-05. Newport, R.I.: August 2005.

Reckner, James R. *Teddy Roosevelt's Great White Fleet.* Annapolis, Md.: Naval Institute Press, 1988.

Seiple, Chris. *The U.S. Military/NGO Relationship in Humanitarian Interventions.* Carlisle Barracks, Pa.: U. S. Army War College, Peacekeeping Institute Center for Strategic Leadership, 1996.

United Nations. *Report and Summary of Main Conclusions.* Regional Workshop on Lessons Learned and Best Practices in the Response to the Indian Ocean Tsunami, Medan, Indonesia, 13–14 June 2005. New York: Office for the Coordination of Humanitarian Affairs, 5 July 2005, available at www.reliefnet.int.

Wylie, J. C. *Military Strategy: A General Theory of Power Control.* New Brunswick, N.J.: Rutgers Univ. Press, 1967.

NEWPAPERS AND JOURNALS

Alexander, Richard. "SECNAV: We Must Think Like Our Enemies." *Newport Navalog,* 16 June 2006.

Allen, Mike. "Flooded Nations Get U.S. Help." *Washington Post,* 27 December 2004.

Arnold, Wayne. "In Areas All But Unreachable, Helicopter Crews Bring Help." *New York Times,* 5 January 2005.

———. "Satisfying Job for the Navy: Saving Lives of Survivors." *New York Times,* 6 January 2005.

Becker, Elizabeth. "U.S. Nearly Triples Tsunami Aid Pledge, to $950 Million." *New York Times,* 10 February 2005.

Bonner, Raymond. "Once-Critical Indonesians Are Grateful to U.S. for Aid." *New York Times,* 7 January 2005.

———. "U.S. Calls Indonesia Deadline for Troop Pullout Reasonable." *New York Times,* 14 January 2005.

Boylan, Peter. "Military to Continue Tsunami Aid." *Honolulu Advertiser,* 11 January 2005.

Brooke, James. "Ad-hoc Planning Goes into Marine Aid Effort." *New York Times,* 6 January 2005.

Brummitt, Chris. "Group Linked to Al Qaeda Starts 'Relief' Work in Aceh." *Washington Times,* 7 January 2005.

———. "Security Threats Won't Limit Efforts." *Boston Globe,* 8 January 2005.

Burgess, Lisa. "Commanders Meeting to Formulate Long-Range Goals in Asian Relief Effort." *European Stars and Stripes,* 8 January 2005.

Choong, William. "U.S. Admiral Expects Better Bilateral Ties." *Singapore Straits Times,* 8 February 2005.

Coutros, Evonne. "A Flying Mission of Mercy." *Bergen County (N.J.) Record,* 19 January 2005.

Demick, Barbara. "U.S. Troops Are Relieved to Be Greeted with Smiles." *Los Angeles Times,* 11 January 2005.

Desjarlais, Orville F., Jr. "Mission of Mercy: Air Force Rescue." *Airman* (April 2005).

Eversley, Melanie. "Relief Groups Pick Up Where U.S. Leaves Off." *USA Today,* 25 January 2005.

Frankel, Rafael D. "U.S. Cargo Plane Brings Lifesaving Supplies to Region." *New York Daily News,* 11 January 2005.

———. "U.S. Copters a Golden Vision—UN: *Mercy* Mission Saluted." *New York Daily News,* 3 January 2005.

———. "U.S. Soldiers Relish Their New Role: Lifesavers." *Boston Globe,* 12 January 2005.

Fuentes, Gidget. "*Mercy* to Revisit Tsunami-Ravaged Area in Spring." NavyTimes.com, 14 February 2006.

Gecker, Jocelyn. "Copter Crash Adds to Tsunami Relief Challenges." *Philadelphia Inquirer,* 10 January 2005.

Gima, Craig. "Hawaii-Based Pacific Command Spearheads Massive Relief Effort." *Honolulu Star-Bulletin,* 3 January 2005.

———. "Pacific Commander Plans to Visit Devastated Region." *Honolulu Star-Bulletin,* 11 January 2005.

———. "Returning Military Gets Counseling First." *Honolulu Star-Bulletin,* 21 January 2005.

Gittler, Juliana. "U.S. Military Begins to Leave Tsunami-Stricken Region." *Pacific Stars and Stripes,* 24 January 2005.

Gomez, Jim. "Aceh Rebels Say Military Broke Truce." *Philadelphia Inquirer,* 22 January 2005.

Goodrich, Robert. "Relief Effort Dwarfs All for Scott Command, General Says." *St. Louis Post-Dispatch,* 7 January 2005.

Graham, Bradley. "U.S. Relief Effort in Asia Not Affecting Iraq Forces." *Washington Post,* 5 January 2005.

Gray, Denis D. "Indonesia Denies U.S. Pilots Use of Airspace; Carrier Leaves." *Washington Times,* 13 January 2005.

———. "Tsunami Horror Hits Close to Home." *Washington Times,* 8 January 2005.

———. "U.S. Choppers Find Remote Survivors: Additional Marines Head to Indo." *Washington Times,* 4 January 2005.

———. "U.S. Military Plans on Long Stay to Airlift Supplies to Coast." *Boston Globe,* 10 January 2005.

———. "U.S. Military Speeds Relief: Air Drops Offer View of Tsunami Destruction." *Washington Times,* 3 January 2005.

"Gunshots Near U.N. Relief Site." *Los Angeles Times,* 9 January 2005.

Hall, Mimi, and Barbara Slavin. "Logistics Delaying Supplies in S. Asia: U.S. Doubles Its Copters for Relief." *USA Today,* 5 January 2005.

Harnden, Toby. "'Time Is the Big Enemy' as Scientists Check the DNA of the Dead." *London Sunday Telegraph,* 9 January 2005.

Harris, Edward. "Fumigation Started to Curb Malaria." *Washington Times,* 15 January 2005.

Hedges, Stephen J. "U.S. Has New Mission at Thai Air Base." *Chicago Tribune,* 6 January 2005.

Herman, Burt. "U.S. Renews Promise to Depart Indonesia." *Washington Times,* 16 January 2005.

Hooper, Craig. "Building a Fleet for Coastal Rescue." *Boston Globe,* 4 January 2005.

Ibison, David, and Jake Lloyd-Smith. "Indonesia Delays U.S. Navy Aid Mission." *London Financial Times,* 10 January 2005.

Ignatius, David. "More Water Bottles, Fewer Bullets." *Washington Post,* 4 January 2005.

"Indonesia's Stature Rises with New Security Pact." *Christian Science Monitor,* 17 January 2006.

"Interview with Gen. Mike Hagee, Commandant, U.S. Marine Corps." *Defense News,* 7 February 2005.

Jaffe, Greg. "U.S. Armed Forces Stretched Further by Aid to Asia." *Wall Street Journal,* 6 January 2005.

Johnson, Tim, and Ron Hutcheson. "Warning Cloud over Relief Effort." *Philadelphia Inquirer,* 11 January 2005.

Kelley, Jeffrey. "Mortuary Affairs Unit Heads to South Asia from Fort Lee." *Richmond Times-Dispatch,* 4 January 2005.

Kessler, Glenn. "White House to Seek $950 Million for Tsunami Aid." *Washington Post,* 10 February 2005.

Kimble, Carrie. "*San Jose* Is a Vital Part of Operation UNIFIED ASSISTANCE." *Navy Supply Corps Newsletter,* May 2005.

King, Neil, Jr., and Donald Greenlees. "Aid Effort in Indonesia Could Lift U.S. Image in Eyes of Muslims." *Wall Street Journal,* 5 January 2005.

Leavitt, Mark S., Jeffrey M. Vorce, and Michael M. Hsu. "For Compassion and Country—UNIFIED ASSISTANCE." U.S. Naval Institute *Proceedings* 131, no. 4 (April 2005).

Leinwand, Donna. "Helicopters Help Relief Operations Gain Momentum: Can Fly Where Bridges, Roads Are Washed Out." *USA Today,* 3 January 2005.

Lekic, Slobodan. "Indonesia, Rebels OK Truce Talks: U.S. Calms Fear over Tsunami Aid." *Washington Times,* 24 January 2005.

"*Lincoln* in Indonesia." *Wall Street Journal,* 4 January 2005.

Malkin, Michelle. "A Salute to the USS *Abraham Lincoln.*" Creaters.com, 5 January 2006.

Margolin, Daren K. "Humanitarian Assistance/Disaster Relief Operations: Combined Support Group Operations in Sri Lanka." *Marine Corps Gazette* (October 2005).

McCammon, Mandy L. "WestPac Express Supports Humanitarian Assistance Mission." Hornblower Marine Services, 10 January 2005.

McNeil, Donald G., Jr. "Hearts, Minds and Flood Relief." *New York Times,* 9 January 2005.

———. "Indonesia Orders Foreign Troops Providing Aid to Leave by March 26." *New York Times,* 13 January 2005.

"Meeting the Enemy." *Baltimore Sun,* 10 January 2005.

"Naval Base Is Spared Damage by Tsunamis." *International Herald Tribune,* 5 January 2005.

"Navy Logistics Supports Operation UNIFIED ASSISTANCE." *Navy Supply Corps Newsletter* (March/April 2005).

"Navy, Marines Deploy to Southeast Asia for Tsunami Disaster Relief." *Inside the Navy,* 10 January 2005.

Norton-Taylor, Richard. "U.S. Island Base Given Warning: Bulletins Sent to Diego Garcia 'Could Have Saved Lives.'" *London Guardian,* 7 January 2005.

"Operation UNIFIED ASSISTANCE: Naval Aviation's Swift Response to the Naval Disaster." *Naval Aviation News* (March–April 2005).

Paddock, Richard C. "Navy Airlifts Supplies to Ravaged Coast." *Los Angeles Times,* 3 January 2005.

"Pentagon Says It's Spending $6 Mln a Day in Tsunami Relief." *Wall Street Journal,* 6 January 2005.

Perlez, Jane. "Indonesia Puts Curbs on Relief in Rebel Areas." *New York Times,* 12 January 2005.

———. "Pace of Global Aid Effort Picks Up on Indonesian Coast." *New York Times,* 4 January 2005.

Pottinger, Matt, and Barry Wain. "U.S. Military Ties to Region Allowed Faster Response." *Asian Wall Street Journal,* 10 January 2005.

Quigley, Samantha L. "High-Speed Vessel Proves Worth during Disaster Relief Mission." American Forces Press Service, 28 January 2005.

Raum, Tom. "U.S. May Boost Image If Efforts Seem Nonpolitical." *Philadelphia Inquirer,* 5 January 2005.

Rhem, Kathleen T. "'Conditions for Transition' Met in Thailand, Sri Lanka." American Forces Press Service, 15 January 2005.

Richter, Paul. "Tsunami Aid Stems Unrest, Boosts U.S. Security, Powell Says." *Los Angeles Times,* 5 January 2005.

Rogers, David. "Lawmakers Expect to Increase Funds for Tsunami Relief." *Wall Street Journal,* 5 January 2005.

Rosenfeld, Isadore. "A New Way to Treat the World." *Washington Post-Parade,* 12 June 2005.

Samuels, Lennox. "Marines Arrive in Sri Lanka." *Dallas Morning News,* 6 January 2005.

Sands, David R. "U.N. Takes Charge of Effort to Help Tsunami Victims." *Washington Times,* 7 January 2005.

Scarborough, Rowan. "Aid Effort Stretches Military Resources." *Washington Times,* 5 January 2005.

Schmitt, Eric. "U.S. and Indonesia May Restore Military Link." *New York Times,* 17 January 2005.

Shane, Scott. "Powell, Praising Relief Effort, Views Sri Lanka Devastation." *New York Times,* 8 January 2005.

Shanker, Thom, and James Brooke. "Tsunami Tests U.S. Forces' Logistics, but Gives Pentagon a Chance to Show a Human Face." *New York Times,* 9 January 2005.

Sherwell, Philip, and Inigo Gilmore. "U.S. Navy Officer Attacks 'Travelling Circus of Aid Workers' for Impeding the Tsunami Relief Effort in Indonesia." *London Sunday Telegraph,* 30 January 2005.

Sipress, Alan, and Ellen Nakashima. "Indonesia Sets Limits on Foreign Relief Role: U.S. Moves Out Carrier; Marines Adjust Plans." *Washington Post,* 13 January 2005.

———, and Glenn Kessler. "A Clear Line between Life and Death." Washington Post Foreign Service, 5 January 2005.

———, and Noor Huda Ismail. "Relief Transcends U.S.-Indonesia Divide: Rights Concerns Sidestepped as Militaries Respond Jointly to Disaster." *Washington Post,* 4 January 2005.

Slavin, Barbara, and Kathy Kiely. "Summit Today on $3 Billion in Aid." *USA Today,* 6 January 2005.

Slavin, Erik. "General: Air Force Shifting Efforts toward Harder-Hit Areas in Indonesia." *Pacific Stars and Stripes,* 23 January 2005.

Spencer, Richard. "U.S. Marines Stage Beach Landings to Bring in Supplies." *London Daily Telegraph,* 11 January 2005.

"Supplemental Request Might Be Delayed to Allow Clear Focus on Tsunami Relief." *National Journal's CongressDailyAM,* 6 January 2005.

Taylor, Guy. "Forensic Teams Sent to Tsunami-Hit Areas." *Washington Times,* 6 January 2005.

———. "Tsunami Relief Runs into Snags." *Washington Times,* 9 January 2005.

Tayler, Letta. "When Relief Fights Belief." *Long Island Newsday,* 23 January 2005.

Unmacht, Eric. "View from U.S. Chopper." *Christian Science Monitor,* 6 January 2005.

"U.S. Displays Generosity, Faces Commitment Test." *USA Today,* 4 January 2005.

"U.S. Military Ending Tsunami Relief Effort: Aid Groups Pledge to Bear Greater Share of Burden as Troops Begin Withdrawal." *Baltimore Sun,* 21 January 2005.

"U.S. Waits to Hear if Troops Needed in Tsunami Area—CNN." *Wall Street Journal,* 28 December 2004.

Vandehei, Jim, and Robin Wright. "Bush Says America Will Lead Global Aid Effort." *Washington Post,* 30 December 2004.

Wallsten, Peter, and Edwin Chen. "Bush Defends U.S. Response to Disaster." *Los Angeles Times,* 30 December 2004.

Welch, William M. "Aid Reaches Remote Areas: Powell: U.S. Help Was Appropriate." *USA Today,* 3 January 2005.

White, Josh. "Wolfowitz Cites Sri Lanka's Progress On Reconstruction: Regional Relief Efforts Called 'Platform' for Future." *Washington Post,* 18 January 2005.

———. "Wolfowitz: U.S. Forces Welcome in Indonesia." *Washington Post,* 15 January 2005.

Wiltrout, Kate. "Local Navy Unit Coordinates Aid Mission." *Norfolk Virginian-Pilot,* 5 January 2005.

Wiseman, Paul. "U.S. Troops Report Successes in Relief Effort: Governments around World Sending Help." *USA Today,* 12 January 2005.

Zimmerman, Fred. "High-Speed Catamaran's Capabilities Put to Use in Relief Effort." *Pacific Stars and Stripes,* 11 January 2005.

———. "U.S. AID Teams Help Military Deliver Relief to Where It's Needed in Disaster Area." *European Stars and Stripes,* 8 January 2005.

———. "U.S. Military Now Has 16,500 Working in South Asia on Tsunami Relief Effort." *Pacific Stars and Stripes,* 10 January 2005.

WEBSITES

edition.cnn.com

gopher.info.usaid.gov

india.indymedia.org

news.bbc.co.uk

news.nationalgeographic.com

seattlepi.nwsource.com

timesofindia.indiatimes.com

usinfo.state.gov

wcatwc.arh.noaa.gov

www.abc.net.au

www.af.mil

www.asiamedia.ucla.edu

www.au.af.mil

www.cpf.navy.mil

www.dod.mil

www.dtic.mil/doctrine

www.guardian.co.uk

www.hinduonnet.com

www.ldeo.columbia.edu

www.pac.dfo-mpo.gc.ca

www.pacom.mil

www.projecthope.org

www.smh.com.au

www.tamilnet.com

www.telegraph.co.uk

www.terradaily.com

www.theage.com.au

www.time.com

www.un.org

www.usatoday.com

www.washingtonpost.com

www.whitehouse.gov

www.house.gov

About the Author

Dr. Elleman received a master of philosophy degree in 1987, an East Asian Certificate in 1988, and his PhD in 1993 at Columbia University. In addition, he completed a master of sciences degree at the London School of Economics in 1985, and a master of arts in national security and strategic studies (with distinction) at the Naval War College, Newport, Rhode Island, in 2004.

Elleman's dissertation research on Sino-Soviet diplomatic relations was conducted in Russia (1988–89), the People's Republic of China (1990–91), Taiwan (1991–92), and Japan (1992–93). Dr. Elleman was a Title VIII Fellow, Hoover Institution, Stanford University, 1993–94, and National Fellow, Hoover Institution, Stanford University, 1994–95. Elleman then taught in the History Department at Texas Christian University. He spent the 1998 calendar year at International Christian University, Tokyo, as a visiting fellow. In 2000, Dr. Elleman moved to the Center for Naval Warfare Studies at the U.S. Naval War College. In 2002–2003 he was a research fellow at the Slavic Research Center, Hokkaido University, and the recipient of a Social Science Research Council grant for advanced research on Japan.

His dissertation was published as *Diplomacy and Deception: The Secret History of Sino-Soviet Diplomatic Relations, 1917–1927* (M. E. Sharpe, 1997). He coedited with Stephen Kotkin *Mongolia in the Twentieth Century: Landlocked Cosmopolitan* (Sharpe, 1999). His other books are *Modern Chinese Warfare, 1795–1989* (Routledge, 2001, translated into Chinese); *Wilson and China: A Revised History of the 1919 Shandong Question* (Sharpe, 2002); *Naval Mutinies of the Twentieth Century: An International Perspective,* edited with Christopher Bell (Frank Cass, 2003, translated into Czech); *Japanese-American Civilian Prisoner Exchanges and Detention Camps, 1941–45* (London: Routledge, 2006); and *Naval Blockade and Seapower: Strategies and Counter-Strategies, 1805–2005,* edited with S.C.M. Paine (Routledge, 2006). An edited book (with S.C.M. Paine), *Naval Coalition Warfare: From the Napoleonic War to Operation Iraqi Freedom,* is forthcoming in 2007. Currently Elleman is writing a history of the Chinese navy and, with S.C.M. Paine, a textbook on Chinese history. A fourth book in the maritime series that began with the 2003 work on naval mutinies is projected, focusing on piracy.

UNIV DUNELM

The Newport Papers

U.S. Naval Strategy in the 1990s: Selected Documents, edited by John B. Hattendorf (no. 27, September 2006).

Reposturing the Force: U.S. Overseas Presence in the Twenty-first Century, edited by Carnes Lord (no. 26, February 2006).

The Regulation of International Coercion: Legal Authorities and Political Constraints, by James P. Terry (no. 25, October 2005).

Naval Power in the Twenty-first Century: A Naval War College Review *Reader,* edited by Peter Dombrowski (no. 24, July 2005).

The Atlantic Crises: Britain, Europe, and Parting from the United States, by William Hopkinson (no. 23, May 2005).

China's Nuclear Force Modernization, edited by Lyle J. Goldstein with Andrew S. Erickson (no. 22, April 2005).

Latin American Security Challenges: A Collaborative Inquiry from North and South, edited by Paul D. Taylor (no. 21, 2004).

Global War Game: Second Series, 1984–1988, by Robert Gile (no. 20, 2004).

The Evolution of the U.S. Navy's Maritime Strategy, 1977–1986, by John Hattendorf (no. 19, 2004).

Military Transformation and the Defense Industry after Next: The Defense Industrial Implications of Network-Centric Warfare, by Peter J. Dombrowski, Eugene Gholz, and Andrew L. Ross (no. 18, 2003).

The Limits of Transformation: Officer Attitudes toward the Revolution in Military Affairs, by Thomas G. Mahnken and James R. FitzSimonds (no. 17, 2003).

The Third Battle: Innovation in the U.S. Navy's Silent Cold War Struggle with Soviet Submarines, by Owen R. Cote, Jr. (no. 16, 2003).

International Law and Naval War: The Effect of Marine Safety and Pollution Conventions during International Armed Conflict, by Dr. Sonja Ann Jozef Boelaert-Suominen (no. 15, December 2000).

Theater Ballistic Missile Defense from the Sea: Issues for the Maritime Component Commander, by Commander Charles C. Swicker, U.S. Navy (no. 14, August 1998).

DURHAM UNIVERSITY LIBRARY
3 0104 01570582 9

Sailing New Seas, by Admiral J. Paul Reason, U.S. Navy, with David G. Freymann (no. 13, March 1998).

What Color Helmet? Reforming Security Council Peacekeeping Mandates, by Myron H. Nordquist (no. 12, August 1997

The International Legal Ramific
Problems and Prospects, by Fran

Durham University
Durham University Library
Please return or renew this item by the last date shown below or the date shown on your Self-Issue receipt.
If this item is requested by another user it may be recalled and a revised due date will be given. The item MUST be returned by the revised due date. Fines are incurred on overdue books.
www.durham.ac.uk/library/

Chaos Theory: The Essentials f
U.S. Air Force (no. 10, Octobe

A Doctrine Reader: The Navies
Spain, by James J. Tritten and V
December 1995).

Physics and Metaphysics of Dete
(no. 8, December 1994).

Mission in the East: The Buildi
by Colonel Mark E. Victorson,

The Burden of Trafalgar: Decisi
the First World War, by Jan S. E

Beyond Mahan: A Proposal for
Colonel Gary W. Anderson, U

Global War Game: The First Fi

The "New" Law of the Sea an
Robertson, Jr. (no. 3, October

Toward a Pax Universalis: A H
1990s, by Lieutenant Colonel
1992).

"Are We Beasts?" Churchill and the Moral Question of World War II "Area Bombing," by Christopher C. Harmon (no. 1, December 1991).

Newport Papers 4, 10, and from 14 on are available online (Acrobat required) at www.nwc.navy.mil/press/npapers/newpaper.htm.